THE AUTHOR OF LIGHT

Did God Reveal His Identity in the Physics of Light?

Dr. Doug Corrigan

The Author of Light – Did God Reveal His Identity in the Physics of Light?
Copyright © 2020 by Douglas Corrigan. All Rights Reserved.

All rights reserved. No part of this book may be reproduced in any form or by any electronic or mechanical means including information storage and retrieval systems, without permission in writing from the author.

Doug Corrigan
Please visit www.ScienceWithDrDoug.com

Printed in the United States of America

First Printing: September 2020

ISBN- 9798691341410

"As a human being, one has been endowed with just enough intelligence to be able to see clearly how utterly inadequate that intelligence is when confronted with what exists."
—ALBERT EINSTEIN

About the Author

Doug has a Ph.D. in Biochemistry and Molecular Biology, a master's degree in Engineering Physics (concentration in Solid State Physics), and a bachelor's degree in Engineering Physics with a minor in Electrical Engineering. As a NASA Graduate Fellow, he worked with NASA on a series of microgravity research studies that flew aboard the Space Shuttle, as well as with the Department of Energy doing research on new materials. He then switched into the life sciences and launched a biotechnology company that developed novel drug-discovery tools. He is also an avid innovator and has won over 30 different awards and licenses for his innovative solutions to problems that span from biotech, to nanoscience, to medicine, to energy technologies.

Doug is married to Amy and is father to three daughters and one son. In his spare time, Doug enjoys composing piano music.

You can contact Doug for speaking engagements or with any questions by emailing:
DrDougCorrigan@gmail.com

You can find more content developed by Dr. Doug Corrigan at:
www.ScienceWithDrDoug.com

Acknowledgements

I would like to thank my friend, Ben Thompson, for his wonderful work in designing the cover of this book, for being a great friend, and for all of the discussions over the years. I want to also thank my wife and the love of my life, Amy, for granting me the time to work on this book and for being a wonderful wife and mother to our children. And I would like to thank my four wonderful children for making life so interesting and funny, and for allowing me to lecture science to you when I drove you to school, when we were gathered at the dinner table, or any other time I could catch your ear.

CONTENTS

PREFACE 1
I. INTRODUCTION 3
II. THE ABSOLUTENESS OF GOD 8
III. THE TIMELESSNESS OF GOD 16
IV. THE OMNIPOTENCE OF GOD 24
V. THE OMNIPRESENCE OF GOD 29
VI. THE OMNISCIENCE OF GOD 33
VII. THE THREE-FOLD NATURE OF GOD – PART I 37
VIII. THE THREE-FOLD NATURE OF GOD – PART II 50
IX. THE THREE-FOLD NATURE OF GOD – PART III 55
X. THE THREE-FOLD NATURE OF GOD -PART IV 60
XI. HOLDING ALL THINGS TOGETHER 64
XII. GOD BECAME MAN 68
XIII. BIRTH, DEATH, AND RESURRECTION 73
XIV. PUTTING IT ALL TOGETHER 82
XV. WHAT DOES THIS MEAN FOR ME? 90

PREFACE

Who created me? How did the universe get here? Why am I here? Why does "existence" exist? What is the purpose of my life? What happens after I die? Is there more to existence than the material world? Is there a God? If so, which God is the true God? Is there any evidence for God? Where do I look for this evidence? Did God leave behind any messages for us? Where do I find these messages, and how do I know if they are true and not just the product of human imagination or dogmas?

Have you ever asked yourself any of these questions? These are very deep questions that have filled the minds, discussions, and writings of philosophers, theologians, and scientists. Even so, these same questions are important to every single person who has lived on this earth.

Have you found the answers to your deepest questions? Perhaps you are satisfied with the answers you have arrived at, but perhaps it is a good time to revisit these questions. It's always a good exercise to reflect and do a sanity check, so to speak. Or, perhaps, you've lived a very busy or difficult life and haven't sat in the quiet long enough to consider these questions. Whatever your situation, I hope you will consider slowing down for a few hours to read this book.

My mind is wired to view the world around me scientifically, which means that my mind is always filled with curiosity. That curiosity started when I was young. Although my curiosity initially centered on the physical aspects of the universe, it quickly branched off into deeper philosophical questions about our existence and about God. During my studies and introspections, a beautiful picture began to emerge: it became increasingly evident that a message was being communicated to us in the physics of the universe, and the purpose of this message was to reveal to us the exact identity of the Creator.

Not only is the Creator's identity revealed in this physics, but also the personal characteristics and actions of the creator. In this book, I will walk you through how the physics governing light, space, time, matter, and energy communicates to us a complete story about our origins, why we are here, and how we should respond. The good news is that you will not need any background in science or physics to understand the content of this book. I present the story in simple concepts that every person can understand. I designed this book for adults and teenagers alike.

If you already know the Creator that I uncover in this book, then I believe the information presented in this book will provide you with a newfound sense of awe and wonder. If you haven't come to know this Creator, then I would only ask that you read this book and consider the science and the message. Who knows, perhaps it will help you to answer some of those nagging questions I first asked.

With that being said, let's start the journey together, shall we?

I. INTRODUCTION

How it all started

If the universe were created by someone, is it reasonable to assume that the creator would leave a signature? I believe that the answer to this question is yes. Leaving a signature behind would serve several purposes.

First, it would authenticate the identity of the true Creator. Almost everything that is made by humans, whether it be a book, a piece of art, a software package, a toothbrush, or a car - is given a type of marking to authenticate the identity of the original creator. Books have authors, art pieces have signatures, and products possess brand names, logos, patents, trademarks, embossings, labels, design features, and serial numbers. These markings are all used to communicate the true identity of the product's creator.

Secondly, a created piece can convey information about the personality and nature of the creator to those who would interact with the created piece. When we study Rondo a Capriccio by Beethoven, or At Eternity's Gate by Vincent van Gogh, we learn something about the personality, character, and nature of the creator of that piece.

Wouldn't it be reasonable to assume that God, the Author of all creation, would do no less? Since God is infinite in wisdom and power, wouldn't it be your expectation that the creation would be filled with so many different types and layers of autographs that these signatures would appear to never end? Certainly, this cannot be proven, but I think that both tenets are reasonable expectations.

After many years of studying physics, quantum physics, biochemistry and molecular biology, I have come to the conclusion that God imprinted His character, nature, and identity into the physical laws that govern light, space, time, and matter. In this book, I will focus mostly on the physics of light because this is where my journey began.

My sincere hope is that you will know the true identity of the Author of light by the end of the book.

My expedition into this topic started as a teenage boy living in New Jersey. As far back as I can remember, I was intensely interested in science. I wanted to know how everything worked. I remember asking my parents and teachers questions that they were not able to answer, and so I read any science book I could get my hands on. I was like a sponge, and my curiosity never seemed to be satisfied. Because my childhood was filled with some grief, I believe that burying my head in science books was my coping mechanism.

I didn't grow up in a religious home, so my knowledge or understanding of God was virtually non-existent. I was a blank slate and had no biases or preconceived notions or ideas. I can't remember my exact age, but I believe I was around 15 years old when some friends of the family had visited our house one day and began to describe how they had been on a quest to understand the truth about God. They explained to us how they explored many different religions, and never really found any truth that they could hang their hat on; however, they communicated to us that there was one faith they had come across that resonated as truth, and they wanted to share some of what they had learned with our family. They told us that this truth was residing in the Bible, which I was completely unfamiliar with at this time.

Around this timeframe, I was studying Albert Einstein's Special Theory of Relativity. I was completely fascinated by the elegance of this theory and how it started with light, and then demonstrated how space, time, and mass are all connected through the physics of light. Light seemed to calibrate every aspect of the universe by acting as a universal standard.

Our friends encouraged us to open the Bible and to start reading. So, I started in Genesis:

> *"In the beginning God created the heavens and the earth."*

I can remember this verse resonating with me in a profoundly impactful way. It drew me in, and I wanted to read more. I kept reading ......

> *"And God said, 'Let there be light,' and there was light."*

The mention of light in this verse especially captivated me because light was the centerpiece of the Special Theory of Relativity. This theory was derived by Einstein by first studying the fundamental properties of light.

Our friends had handed me an audio tape and asked me to listen to it. I can't remember the name of the person speaking on the audiotape, but my recollection was that it may have been a pastor of a church. On this tape, he discussed how certain aspects of the Special Theory of Relativity could be lined up with the characteristics of God.

This initial seed intrigued me and started me on a journey of fully understanding the physics of light. Various concepts presented in this book didn't come into my thinking until after going to college and studying physics in great depth. That being said, you will not need any background in science or physics to understand the content in this book. I believe that most teenagers and adults can understand the concepts presented in this book.

As you will discover along with me in this book, I believe that the God described in the Bible designed the laws that govern light in such a way to communicate His nature and identity to us. God inserted a hidden code in the physics of light that would not be fully understood until the modern era. After thinking about this problem deeply over the last 30+ years, I have come to the belief that the identity of the true God is residing in the physics of light. His identity is also expressed in many other sources, but the focus of this book is on the physics of light and how light is related to the physics of space, time, matter and energy. In the rest of this book, I will break down how different aspects of the physics of light directly correlate with characteristics of the God described in the Bible. In each chapter, I will introduce a new property or identifying feature.

I think it is important to point out that I in no way believe that God is somehow residing in, or encompassed by, a physical beam of light. Rather, I am making the case that the physical behavior of light is a metaphor that parallels His character and identity. Light is an image of God, which is quite apropos given the purpose of light. This is profoundly noteworthy, since God frequently refers to Himself as light throughout the scripture:

> *I John 1:5* – *"This is the message we have heard from him and declare to you: God is light; in him there is no darkness at all."*

> *John 8:12 – "When Jesus spoke again to the people, he said, 'I am the light of the world. Whoever follows me will never walk in darkness, but will have the light of life.'"*

In a strange way, Albert Einstein was a piece of the puzzle that led me to understanding God. From what I know about him, I do not believe he saw these connections because they are not evident in his writings. Physicists have learned a great deal more about the nature of light since his passing, and I will discuss those aspects of light and how they relate to God in this book as well.

With that backdrop, let me now take you on a journey that I believe will amaze you and bring you closer to the Author of all creation.

II. THE ABSOLUTENESS OF GOD

A strange property of light

I will start with Albert Einstein's Special Theory of Relativity, because this is where I began my voyage as a teenager. Although this theory can be quite complex and difficult to understand, I will keep the discussion simple to provide you with just enough of a basic understanding to grasp the theory's general concepts.

If you have never been exposed to any of these concepts before, then what I am about to share with you will sound very strange, but that's ok. You may need to read this chapter twice, and that's o.k. You may need to read this chapter three, four, five, six - or even ten times - and that's o.k. Most people have this initial reaction because the conclusions stemming from the Special Theory of Relativity are counterintuitive and fly in the face of our notions about how the world should work. I am going to remove the math from the discussion so that we can focus on the concepts. Please stay with me, because the punchline is extraordinarily profound.

The Special Theory of Relativity was first presented by Albert Einstein in 1905. This theory is fundamentally about how time, space, and matter and energy are all related through the absolute standard of light. Even though the conclusions drawn from this theory are somewhat bizarre and unusual, this theory has been proven experimentally through thousands of experiments.

The Special Theory of Relativity is derived from two key postulates. Einstein was influenced by the previous work of other physicists, such as Hendrik Lorentz and Henri Poincaré. The first postulate is that the laws of physics are identical in all frames of reference (for frames of reference that are moving at constant speed). For example, a scientist conducting experiments in a spaceship moving at a constant speed of 100,000 miles per hour would find that the laws of physics are the same as a scientist conducting experiments in a spaceship moving at a constant speed of 10 miles per hour. Essentially, there are no privileged reference frames. A reference frame is just a fancy way of saying "your point of view." There is no reason to think that the laws of physics should change if you are in a car moving at 80 mph, or in a spaceship moving at 20,000 mph.

The second postulate is that the speed of light (186,282 miles per second) is identical for all observers. This second postulate is the key that unlocks every single conclusion that naturally follows from this theory. Einstein recognized that the speed of light was a universal, absolute standard. In a way, it is ironic that a theory with the word "relativity" in its title is based off an "absolute" standard, i.e., the speed of light.

Einstein assumed that the speed of light is an absolute, which means that the speed of light does not change with respect to the relative speed of an outside observer. At first, this might not sound that strange, but it is completely counterintuitive to our understanding of how everyday objects work.

Let me explain. For example, if you were standing on the ground and measured the speed of a car going by at 60 mph, then the speed of that car relative to you (stationary observer) is 60 mph. But what if you are in a car traveling alongside this other car, moving in the same direction at the same speed, 60 mph? In this case, what is the speed of the other car relative to you? The answer is 60 mph - 60 mph = 0 mph. The speed of the other car is **_relative_** to the speed of your car.

This answer would be verified by the fact that as you looked out of your window, you would observe that the other car is traveling at exactly the same rate as you. Because you are both traveling in the same direction at exactly 60 mph, the other car would neither be inching ahead of you, nor falling behind. Since you would be traveling in lockstep with this car, this car's relative speed with respect to your car (the observer) is zero. Assuming both drivers were great drivers and the cars weren't bouncing up and down, you could open your window and seamlessly hand the passenger in the other car a drink.

Now, if you were in deep outer space with no visible landmarks, you would look at the other car and think that you were both stationary (not moving at all). In fact, if another external object is not available as a reference to measure your speed (like a street sign or a tree), it's impossible to determine if you're moving at all! In this way, we see that the concept of speed is relative to something else.

Let's look at another example that demonstrates how speed is relative for normal objects. Let's say that you are capable of throwing a ball at 50 mph. If you were to stand in the back of a pickup truck moving at 60 mph, and threw the ball at 50 mph in the same direction that the truck was moving, the ball would move (with respect to someone stationary on the ground) at a speed of 60 mph + 50 mph = 110 mph. This is due to the fact that the ball in your hand was already moving at 60 mph with respect to the ground before you decided to throw it. Then, your arm gave enough energy to the ball to make it move an additional 50 mph faster; thus, the final speed of the ball is the combination of the speed of the truck plus the additional speed that you imparted to the ball before it left your hand.

With everyday objects, speed is ***relative***. The speed of another object is based on your relative speed with respect to that object. As in the two examples discussed above, speeds add and subtract. Basically, with everyday objects, the speed of one object completely depends on the speed of another. Ordinary objects that we are accustomed to do not possess such a thing as absolute speed. A car moving at 60 mph with respect to the ground will appear to be traveling at different speeds with respect to other observers, depending on the particular motion of the observer.

Einstein turned this notion on its head with light. Einstein started with the assumption that light did not behave this way. Einstein stated that the speed of light is always measured as 186,282 miles per second, regardless of the observer's speed. Einstein's premise was that the speed of light is absolute, and is measured by all observers, regardless of their speed, as the same number: 186,282 miles per second. The speed of light never changes, and it is not dependent on the speed of the observer who is attempting to measure the speed of light.

This is bizarre. This means that if you were in a spaceship moving at 185,282 miles per second (1,000 miles per second slower than the speed of light) alongside a light beam moving in the same direction, you would **not** measure the light beam moving ahead of you at 1,000 miles per second (186,282 miles per second - 185,282 miles per second). Even though your spaceship is moving at 185,282 mps, the light beam would still be moving away from you at 186,282 mps! No matter how fast your ship was traveling, it would never catch up to the light beam because the light beam would always be moving away from the ship at the full speed of light!

This would be like trying to catch up to a car ahead of you which is moving at 60 mph. To your amazement and bewilderment, you would find that the car moving at 60 mph would continue to move away from you at 60 mph, even as you continued to accelerate to 60 mph. No matter how hard you pushed on the gas pedal, the car in front of you would continue to move away from you at 60 mph, even though the person in this car isn't accelerating the car to compensate for your increased speed. Einstein stipulated that light functions like this supernatural car in front of you.

Even if you could muster up enough energy to get up to 99.99999999% the speed of light, a light beam would still move away from your spaceship at the speed of light! (we will see later why it's impossible for an object with mass to actually reach 100% the speed of light). Even more unusual is the fact that a person standing on the ground would measure this same beam of light moving precisely at the speed of light, 186,282 miles per second. How can two different observers, both moving at different speeds, measure the same beam of light moving at the same speed?

In our way of thinking, the only way this could possibly be true is if light were somehow aware of our speed and adjusted its speed in response. For example, if we accelerated our spaceship to catch up with the beam of light, perhaps the light beam would somehow increase its speed by the same amount so that it would always be moving away from us at the same speed. As you will see, this is not what light does.

We know this by the following thought experiment. Consider a single light beam and two observers. The first observer is stationary on the ground, and the second observer is traveling in a spaceship. As the second observer in the spaceship increased its speed to catch up with the light beam, the light beam would still move away from the spaceship at the speed of light. However, the stationary observer would also measure this same light beam to be traveling at the speed of light. If the light beam had increased its speed to get away from the person traveling in the spaceship, then the stationary observer would measure the speed of the light beam as a number greater than the speed of light. But since the speed of light is the same for all observers regardless of their reference frame, this cannot be the case. Both observers must measure the speed of light to be the same number, even though one observer is moving, and the other is not. Obviously, something else is going on here that we do not understand.

If two cars traveling at 60 mph collided in a head-on collision, they would collide at 120 mph. This is because the relative speeds of the cars add together. This is not so with light. Let's say that a spaceship was moving at 50% the speed of light towards a light beam moving in the opposite direction towards the spaceship. In this case, the speed of the light beam with respect to the spaceship would be the speed of light, not 1.5 times the speed of light. If the spaceship were to travel at 99.99999999% the speed of light, the light beam headed straight for the spaceship would still be measured as moving at the speed of light, not ~2 times the speed of light!

No matter how fast the spaceship moves, or in what direction the spaceship moves – the person on the spaceship will always measure the light beam moving at 186,282 miles per second.

Somehow, the light beam is not dependent on anything else but itself. The light beam is not relative to anything else. The speed of light is absolutely absolute.

This is the assumption Einstein started with, and it was proven through many experiments. As we will see in subsequent chapters, this one assumption about the absolute nature of the speed of light resulted in one of the most profound discoveries of the 20th century: that for the speed of light to be absolute, everything else, including the rate at which time flows, is relative! Time changes. Mass changes. Length changes. We think of time, mass, and length as absolutes, but they are actually relative quantities that change. Light is absolute, and everything else is relative with respect to the absoluteness of light. Light calibrates everything else.

We now know the first unique property of light:

Light is absolute.

If we relate this distinctive property of light to the God described in the Bible, the picture becomes very clear. The God presented in the Bible is absolute. In every single aspect, God is absolute. God is dependent on nothing else. God never changes. There is one God who doesn't change. Regardless of what reference frame you view God from, God is the same. God is the source of absolute truth. God is absolute truth. In direct opposition to the philosophy of relativism, God's truth is not relative to anything or anyone else; instead, everything and everyone else is relative to His absolute truth.

Everything in the universe is relative to God's absolute standard, just like every physical property in the universe (space, time, matter, and energy) is relative to the absolute standard of light (more on this later).

Additionally, God is the same to all people regardless of who they are or their reference frame. Regardless of gender, cultural background, social status, or race, God is precisely the same. God does not change over time, and God does not change from person to person.

The relationship we see between light and the physical universe is the same relationship we see between God and people. This attribute of "absoluteness" is a unique identifier of the God presented in the Bible. In many religions, God is shaped in response to the person's desires. Many people want to worship God, but only a God that is acceptable or relevant to their way of thinking. Voltaire stated, "God made man in His image, and man returned the favor." Man wants a

God that he can control. This view is evident in the majority view of modern society, which is the notion that absolute truth does not exist. However, if light is anything like God, then there is an absolute standard that everyone and everything is subject to. Here we see the first parallel between the physics of light and the attributes of God:

God is Absolute ←→ Light is Absolute

> *Malachi 4:6 - "For I, the Lord, do not change."*
>
> *Deuteronomy 6:4 - "The LORD our God, the LORD is one."*
>
> *Exodus 3:14 - "God said to Moses, "I AM who I AM."*
>
> *Psalm 119:160 - "All your words are true; all your righteous laws are eternal."*
>
> *John 1:1 - "In the beginning was the Word, and the Word was with God, and the Word was God."*
>
> *John 14:6 - "Jesus answered, 'I am the way and the truth and the life. No one comes to the Father except through me.'"*

III. THE TIMELESSNESS OF GOD

Time is not what you think

In the previous chapter, Einstein's Special Theory of Relativity was introduced. We were introduced to the concept that the speed of light is absolute, and this was related to God being the absolute source of truth.

This absolute property of light leads to a series of mind-boggling conclusions. We will look at these ramifications in this chapter and the next few chapters. We will demonstrate how each of these conclusions directly relate to specific characteristics of the God.

Einstein discovered that if the speed of light is an absolute constant, then time, space, and matter are relative. Not only is the speed of light absolute, it also calibrates everything of importance in the universe. In this chapter, we will focus on the relativity of time.

Wait, did I just use the phrase "the relativity of time"? This choice of words is strange because we normally think of time as an absolute. Due to our experience with the constant ticking of a clock, we are under the impression that time moves at the same rate for everyone, and that time is governed by some external universal clock that is the same for everyone. How can time be relative?

To understand this concept, let's use a simple thought experiment involving a laser beam which is reflected from a mirror on a train (see figure below). A light beam leaves the laser, travels to the mirror, is reflected from the mirror, and then travels back to the laser. To an observer sitting on the train, the light beam moves straight up and down in a vertical line. The reference frame from the perspective of the person on the train is illustrated in the top graphic.

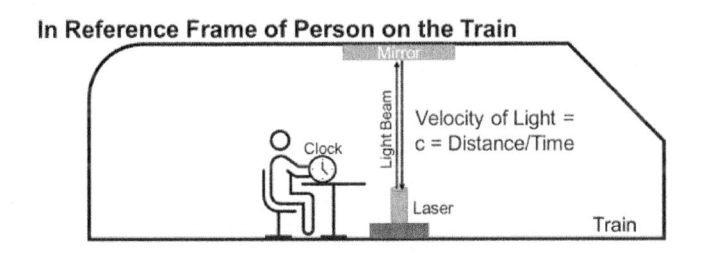

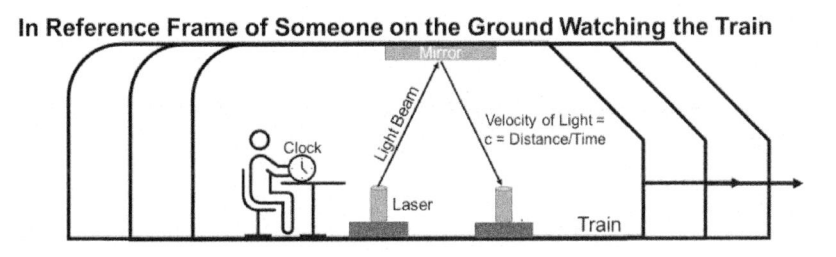

Thought Experiment on Light

In Reference Frame of Person on the Train

Velocity of Light = c = Distance/Time

In Reference Frame of Someone on the Ground Watching the Train

Velocity of Light = c = Distance/Time

From the reference frame of a person standing on the ground watching the train go by, the light beam appears to trace out a zig zag path as it is reflected from the top mirror. The reference frame from the perspective of the person standing on the ground is illustrated in the bottom graphic.

Now, if the person sitting on the train wants to measure the speed of light, they will measure the distance that the light travels along the vertical line between the laser and the top mirror, and from the mirror back to the laser (total distance traveled). Then, they will divide that total distance by the amount of time it takes for the light beam to bounce back and forth. This is because speed is calculated as the distance something travels in a given amount of time.

However, if the person standing on the ground attempts to peer into the train (perhaps through a window) and measure the speed of light, they will measure the distance that the light travels along the zig zag path and divide that by the time it takes for the light beam to travel up and then back down to the laser. The distance that the zig zag pattern traces out is longer than the vertical distance (just like the hypotenuse of a right triangle is longer than either side of the triangle), and so the person on the ground calculates the speed of light to be greater than the speed that someone on the train would measure.

To reiterate, the distance the light appears to travel to the person on the ground is a greater distance than the distance measured by the person on the train. The person on the train measures the distance as the distance between the laser and the mirror, and the person on the ground measures the distance based on a hypotenuse of a triangle. Even if you have trouble remembering this property of right triangles from geometry, you can look at the figure and see that the distance is greater for the person viewing the light beam from the ground.

A greater distance divided by the same amount of time will result in a greater speed, since:

Speed = Distance / Time

Based on what we learned in Chapter 2, do you see a problem? Yes, that's right - this violates the principle that the speed of light is the same to all observers. How can both observers measure a different speed for the speed of light when the speed of light is absolute and isn't dependent on the reference frame of the observer? When the person on the train and the person on the ground compare notes, they will realize that the person on the ground measured a higher speed for the speed of light. We seem to have a major problem.

Einstein realized that the only way for both observers to measure the same speed for light is for the time measurement to change in proportion to the length change. Since speed is distance divided by time, a longer distance divided by a longer time gives the same result as a smaller distance divided by a correspondingly smaller time. If each observer measures a different time so that the ratio stays the same, then the speed of light remains the same. This is illustrated by the fact that 2 divided by 2 is the same as 5 divided by 5. Both ratios result in the number 1. Essentially,

Distance/Time = Distance/Time

So, for the speed of light to remain the same to both observers, the time that the person on the train measures must be less than the amount of time that the person on the ground measures. For the person on the ground, it would have taken longer for the light beam to travel from the laser and return back to its starting point after reflecting off the mirror.

How can two different observers measure different durations for the same event? I think you will agree that the only way this can occur is if time flows at a different rate for each observer! The clock of the person on the train must move slower when compared to the clock of the person on the ground!

If the train moves faster, the zig zag stretches out further, and the total distance traveled by the light beam according to a stationary observer on the ground increases. This translates to the clock moving even slower on the train compared to a clock on the ground.

If you do the math, you find that as the train approaches the speed of light, the clock runs continually slower. If the train could reach the speed of light, time would come to a complete standstill on the train. If the train could reach the speed of light, the person on the train would essentially be timeless. If the train were traveling at the speed of light, an infinite amount of time could elapse for someone standing on the ground, while absolutely no time elapsed for the person on the train.

Thinking about this a bit more, we can conclude that the person on a train that is traveling at the speed of light experiences timelessness! Essentially, all points in time are collapsed to one point in time. Time isn't moving a single millisecond for the person on the train, while the world outside of the train is moving through time!

This sounds crazy! No one believed Einstein's theory when it was first released. After careful experiments were designed and performed, the data conclusively demonstrated that Einstein's theory was completely correct!

As someone moves faster, their clock slows down. The person who is in motion does not realize this slowdown of time because the clock in their head also slows down! To the person who is moving in the train, nothing has changed. It seems as if it is business as usual. The clock of the moving person only appears to be moving slower when observed by someone who is not moving along with the clock! This means that time is not an absolute quantity. Time is completely relative.

It is important to point out that this time difference is extremely small for speeds that we normally move at in a car or in a plane. As an example, the clock of someone moving at 60 miles per hour runs at 0.999999999999996 times the rate of a clock which is stationary in a person's home. A clock in an airplane moving at 500 mph runs at 0.99999999999972 time the rate of clock on the ground. These small time differences are imperceptible in most situations.

However, these miniscule differences in time are important and necessary for GPS satellites. In fact, the GPS in your phone functions based on this math due to the fact that the clocks on GPS satellites run at a different rate than the clocks here on earth. This time difference is extremely small, as the clock on the GPS satellite runs slower by a rate of 7 microseconds (7 millionths of a second) every day due to Special Relativity. This is due to the speed of the satellite as it orbits the earth, which is 14,000 km/hr. This small time difference would result in a large miscalculation of distances here on earth if it is not accounted for.

(Note: there is also a time difference on satellite clocks due to General Relativity, which was formulated by Einstein after Special Relativity was introduced. In General Relativity, Einstein showed that gravity also affects the rates at which clocks move.)

In contrast, at speeds that are a significant fraction of the speed of light, these differences are more pronounced. For example, the clock in a spaceship moving at 50% the speed of light runs at 86.6% the rate of a clock on earth (i.e., the clock runs 13.4% slower). This means that if this spaceship were to travel at this speed for 86.6 years and then return to earth, 100 years would have passed on earth. This space traveler would think that they traveled ~14 years into the future. This appearance of traveling into the future by the space traveler is due to the fact that they experienced 86 years according to their clock, while everyone else on earth experienced 100 years. As another example, if this spaceship traveled at 99% the speed of light for the same 86.6 years, 613.89 years would have passed on the earth. This space traveler would feel like they traveled ~527 years into the future.

Again, this sounds impossible, but this relativity of time based on the absolute speed of light has been proven to many decimal points in countless experiments. Notice that it is possible to travel into the future based on the relative difference in the rates of clocks, but it is impossible to travel into the past. If you want to travel into the future, you will need a spaceship that moves extremely fast. As of today, we do not have spaceships that can move fast enough to travel into the future any appreciable amount.

If someone were actually able to travel at the speed of light, their clock would move infinitely slow. In other words, their clock would be completely stopped while clocks on the earth were moving at their normal rate. This space traveler could travel an infinite number of years into the future and feel like absolutely no time passed! Effectively, if someone were able to reach the speed of light, they would experience all points in time simultaneously. (we will see later why it's impossible for something with mass to reach the speed of light.)

This property can be related to the God described in the Bible, who claimed repeatedly in the scriptures that He is "light." We know from the scriptures that God is "timeless" in that God is not confined or constrained by time. God is not linearly progressing through time. We know that God lives outside of time and that time is a created thing, just like space, matter, and energy. God experiences all points in time (past, present, and future) simultaneously, just like light. God knows the end from the beginning.

Again, we see the peculiar properties of light mapping to the characteristics attributed to an infinite God:

God is Timeless ←→ Light is Timeless

> *Isaiah 57:15 - "For thus says the One who is high and lifted up, who inhabits eternity, whose name is Holy."*
>
> *Psalm 90:4 - "For a thousand years in your sight are but as yesterday when it is past, or as a watch in the night."*
>
> *2 Peter 3:8 - "But do not overlook this one fact, beloved, that with the Lord one day is as a thousand years, and a thousand years as one day."*
>
> *Isaiah 46:10-11 - "I make known the end from the beginning, from ancient times, what is still to come."*

IV. THE OMNIPOTENCE OF GOD

Impossible energy

We are peeling back the layers of light to evaluate the hypothesis that the physics of light is encoded with properties that perfectly parallel the characteristic of the God described in the Bible. This notion is in alignment with the fact that the Bible refers to God as "light" throughout the scripture. From the small amount we have learned so far, it appears that God may have imparted light with a list of curious and unusual physical behaviors to communicate to us His precise identity and nature.

So far, we have seen that the physics of light expresses both the absoluteness and timelessness of God. Both of the characteristics of God are deeply embedded in the unique physical principles that govern the physics light. First, we saw that the absoluteness of God paralleled the absoluteness of the speed of light. Second, the timeless nature of God was revealed by turning our attention to a thought experiment involving a light beam bouncing off of a mirror in a moving train. From this thought experiment, we demonstrated that clocks in motion slow down, and that clocks moving at the speed of light completely stop. Therefore, anything moving at the speed of light does not experience the passage of time. This slowing down of time was one consequence of Einstein's Special Theory of Relativity, which was derived from the initial observation that the speed of light is absolute. This timeless property of light was related to God because God is timeless, as He can experience the past, present, and future simultaneously.

Let us now turn our attention to another consequence of the Special Theory of Relativity:

It is impossible for any object to accelerate to the speed of light due to the fact that this would require an infinite amount of energy.

Why is this? Well, in the Special Theory of Relativity, the mass of an object increases as its velocity increases. This is another natural corollary that follows from the initial assumption that the speed of light is the same in all reference frames (when combined with conservation laws). All objects start at rest with a certain "rest mass." For example, an object might have a rest mass of 1000 grams. This is the type of mass that we are all familiar with.

As an object, such as a spaceship, accelerates to higher speeds, its mass increases. This increased mass is called the "relativistic mass." If an object possesses a greater mass, the amount of energy required to accelerate the object to a higher speed is also greater. It requires ten times more energy to accelerate a 20,000 kilogram truck from 50 mph to 60 mph than it does to accelerate a 2,000 kilogram car.

It can be shown that mass increases by the same factor that governs the slowing down of clocks on the spaceship. So, for example, if the speed is such that the rate of the clock is reduced by a factor of two, then the mass of the spaceship increases by a factor of two.

So, in a sense, mass is relative. It is important to keep in mind that to an observer who is traveling in the spaceship, nothing has changed. Just like the clocks, this change in the mass is with respect to an observer in a different frame of reference. The clocks on the ship appear to be running normally to someone on the ship, and the mass of the ship appears to be normal as well. However, to someone who is at rest and observing the spaceship from afar, it appears that the clocks on the ship are moving slower, and the mass of the ship is increased.

These changes in mass are very small at the velocities we are accustomed to with everyday objects, like cars and airplanes. The increase in mass of a car moving at 60 mph would be imperceptible. In order for these differences to become noticeable, the object needs to move at some speed that is a considerable fraction of the speed of light.

As the spaceship accelerates, its mass increases. As the mass increases, a greater amount of thrust is required to accelerate the spaceship to higher velocities. And as the spaceship gains more speed, it gains more mass, and it requires even more thrust to reach the next velocity. It can be shown mathematically that if a spaceship were to reach the speed of light, its relativistic mass would be infinitely large! In the graph below, you can see this effect. The mass starts out at 1000 grams. As the object's speed initially increases, the mass increases, but it remains very close to 1000 grams. However, as the speed of the object approaches 99% of the speed of light and above, the mass shoots off towards infinity. Special Relativity stipulates that if the speed were exactly equal to the speed of light, this mass would be infinite.

Relativistic Mass of Object Versus Speed

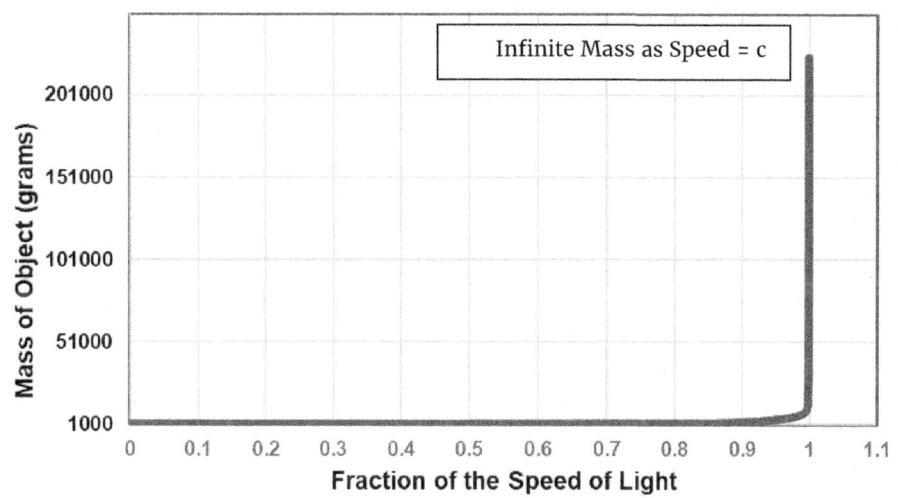

[Graph showing mass of object (grams) on y-axis from 1000 to 201000, versus fraction of the speed of light on x-axis from 0 to 1.1. The curve remains near zero until approximately 0.9, then rises sharply toward infinity as the speed approaches c. Label: "Infinite Mass as Speed = c"]

Therefore, it would require an infinite amount of energy for any object to accelerate to the speed of light. Because we do not have access to an infinite amount of energy, no object in the known universe can accelerate to the exact speed of light.

Even if you attempted to accelerate an object that is incredibly small, like an electron, it would still require an infinite amount of energy. The closest we have come to accelerating subatomic particles to close to the speed of light is at the Large Hadron Collider. The Large Hadron Collider uses 7 Teraelectron volts of energy to accelerate subatomic particles to 99.9999991% the speed of light. To accelerate those particles that last little bit would require all of the energy in the known universe, plus a nice big helping of an infinite amount of energy on top of that!

Any object that starts with a non-zero mass at rest will have an infinite amount of mass and energy if it ever reached the speed of light! Light is the fastest thing in the universe, and no physical object can ever reach this speed. If you want to travel at the speed of light, you need an infinite amount of energy! Therefore, not only is light the absolute standard by which all other things must be measured, it is also completely out of our reach. And remember, even if you could conjure up enough energy to reach 99.9999999999999999999 percent the speed of light, a beam of light would still whiz past you at exactly the speed of light. You can't even begin to catch it!

God went through great lengths in the scripture to relate Himself to light. If God imparted His characteristics into the physics of light, then we can conclude that one of the attributes of God must be infinite power. We know that God is omnipotent, or infinite in power and strength. Nothing is out of God's reach or capabilities. We also know that it is impossible for any person to become God, regardless of how hard they try --just like it is impossible to accelerate any object to the speed of light. God is God, and no person on earth can come close to approaching even a miniscule fraction of His greatness, even if they had access to all of the energy in the universe.

God is Omnipotent ←→ Infinite Power is Required to Reach the Speed of Light

> *Jeremiah 32:17 - "Ah, Sovereign LORD, you have made the heavens and the earth by your great power and outstretched arm. Nothing is too hard for you."*
>
> *Job 11: 7-11 - "Can you fathom the mysteries of God? Can you probe the limits of the Almighty? They are higher than the heavens above—what can you do?*
>
> *Matthew 19:26 - "But Jesus beheld them, and said unto them, with men this is impossible; but with God all things are possible."*

V. THE OMNIPRESENCE OF GOD

Can we hide from God?

In the previous chapters, we discussed the absolute nature of light, and how this property of light resulted in a number of properties that perfectly parallel the attributes of the God described in the Bible. In particular, we demonstrated that the physics of light teaches us that God is absolute, timeless, and omnipotent. God infused light with characteristics that reveal who He is. Now, we will look at another property of light and try to discover what it tells us about God.

In Einstein's Theory of Special Relativity, the speed of light is absolute. From this one principle, a great deal of math flows, and this math reveals some startling conclusions. This math tells us that clocks slow down and that mass increases as an object approaches the speed of light, and that if the object could reach the speed of light (which is impossible), its clock would completely stop (i.e., it would become timeless) and the mass of the object would become infinite (it would have infinite energy). Each one of these bizarre properties of light precisely corresponded with the unique attributes associated with the God of the Bible.

There is one other thing that this math reveals to us:

Length is not an absolute quantity. Length is relative.

As a spaceship travels faster, the distance between objects located outside of the spaceship progressively shortens in the direction of travel. The measurement of length outside of the spaceship contracts in the direction of travel. Conversely, an observer on the ground watching the spaceship zoom by would measure the length of the spaceship as less than its actual length.

As objects approach the speed of light, their length contracts dramatically (when viewed by a stationary observer). If the object could reach the speed of light, its length in the direction of travel would become infinitesimally small. Conversely, objects outside of the spaceship appear to be shorter in length when viewed by the person on the spaceship. Just like time and mass, the person on the ship feels normal. The person on the spaceship doesn't feel like length has contracted on the spaceship. However, if the space traveler peers out of the window, a ruler that is stationary outside of the spaceship would appear to have a reduced length. This principle is illustrated below. A ruler that is positioned on the ground outside of the spaceship would appear to be normal length when viewed by a person standing on the ground. To the space traveler, this ruler would appear to be contracted. As the ship accelerated and approached the speed of light (which is impossible for reasons described in the previous chapter), the ruler's length would shrink to zero!

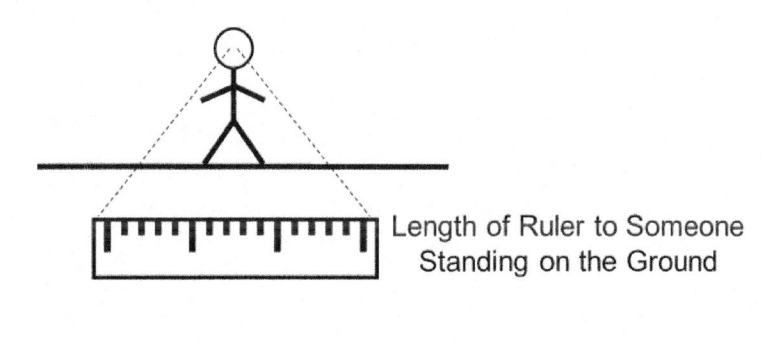

Length Contraction

Length of Ruler to Someone Standing on the Ground

Length of the Same Ruler to Someone Moving in a Spaceship

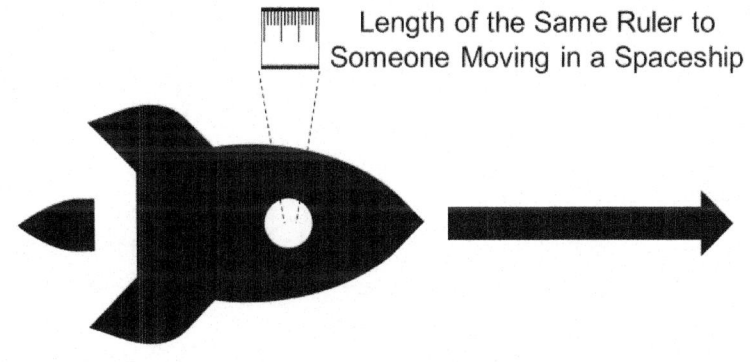

In fact, much more than the ruler would shrink to zero length. If you were on a spaceship traveling at the speed of light, the entire universe would shrink down to an infinitesimally thin two-dimensional plane, much like a piece of paper. This two-dimensional plane would be infinitesimally thin and perpendicular to the direction of travel. One way to think about this is as follows: if you could travel at the speed of light, there would be no distance to travel. In other words, the beginning and the end of your path are at the same point! Also, your clock is completely stopped (see previous chapter on this topic), so absolutely no time passes for you.

Now, if you turn your ship in a different direction, the universe will collapse to an infinitesimally thin two-dimensional plane in that direction. What this means is that you have access to the entire volume of the universe instantaneously. Since time is stopped for you, you would have access to the full volume of the universe at one moment in time. In other words, you would be "omnipresent." Light has an omnipresent-like characteristic to it because length contracts to zero in the direction of travel, and time stands still.

God reveals to us in the scripture that He is light. Our hypothesis throughout this book is that God embedded a hidden code into the physics of light that reveals His identity and nature. If that is true, then what does this property of light tell us about God?

I believe that this property of length contraction demonstrates to us that God is omnipresent. God is everywhere at once! God is not confined by space.

God is Omnipresent ←→ Light is Omnipresent

> *Jeremiah 23:23-24 – "'Am I only a God nearby,' declares the Lord, 'and not a God far away? Can anyone hide in secret places so that I cannot see him' declares the Lord. 'Do not I fill heaven and earth?' declares the Lord"*
>
> *Psalm 139:7-10 - "Where shall I go from your Spirit? Or where shall I flee from your presence? If I ascend to heaven, you are there! If I make my bed in Sheol, you are there! If I take the wings of the morning and dwell in the uttermost parts of the sea, even there your hand shall lead me, and your right hand shall hold me."*

VI. THE OMNISCIENCE OF GOD

Unbounded knowledge and wisdom

Now, we turn our attention to God's omniscience: the fact that God is all knowing and possesses an infinite amount of wisdom and knowledge. This chapter is short, but it's one of my favorite parallels between light and God.

By now, you probably will not be too surprised to find out that light possesses the intrinsic capability to transmit an infinite amount of information on one light beam.

You did not read that statement incorrectly.

Listen to what Mario Krenn, Professor at the University of Vienna and the Institute for Quantum Optics and Quantum Information stated:

> *"The Orbital Angular Momentum OAM of light is theoretically unbounded, meaning that one has, in theory, an unlimited amount of different distinguishable states in which light can be encoded."*

In simple terms, this means that the physics of light allows for encoding a single light beam with an infinite amount of information, and this information would be virtually instantaneously transmitted at the speed of light. There is no other physical medium in which this is possible. This is only theoretical, because we have no way to test such a proposition simply because we do not have access to storing or transmitting an infinite amount of information, but the physics of light would allow for this to occur if it were practical. A single light beam has the capacity to transmit an infinite number of conversations at once.

You may be familiar with the fact that fiber optic cables are used to transmit high volumes of data in a short period of time by taking advantage of the physics of light. In reality, current technology only takes advantage of an infinitesimally small portion of the data-handling capacity of light, and even this infinitesimally small use of the handling capacity of light is staggering. The restriction to accessing the infinite data-handling capacity of light is due to the limitations of our technology, not the limitations of light.

In addition, this property of light allows for absolutely private or secret communication between two people. The orbital angular momentum (OAM) of light photons can be used in cryptographic quantum communication. In this form of communication, a secret "key" is constructed from a string of spinning photons (photons are "particles" of light). This key is shared between two individuals, and this key protects data shared over the link. The laws of quantum physics and the properties of light render any attempt by an eavesdropper to intercept the key as futile. Any attempt to measure the key destroys the key. This type of quantum communication over light is unbreakable.

Therefore, a single light beam is capable of delivering an infinite amount of information whereby only the person or persons with the key have access to this information.

We know from the scriptures that God is infinite in wisdom and understanding (omniscient). We also know that no man can peer into the mind of God, and that His ways are past finding out.

God Is Omniscient ←→ Light Can Transmit an Infinite Amount of Information

God's Mind is Concealed ←→ Light's Communication is Secret and Unbreakable

> *Psalm 147:5 - "Great is our Lord, and mighty in power; his understanding has no limit."*
>
> *Isaiah 40:28 - "Do you not know? Have you not heard? The Lord is the everlasting God, the Creator of the ends of the earth, He will not grow tired or weary and his understanding no one can fathom."*
>
> *Isaiah 55:8-9 – "For my thoughts are not your thoughts, neither are your ways my ways, saith the LORD. For as the heavens are higher than the earth, so are my ways higher than your ways, and my thoughts than your thoughts."*
>
> *Romans 11:33 - "O the depth of the riches both of the wisdom and knowledge of God! how unsearchable are his judgments, and his ways past finding out!"*

> 1 Corinthians 2:9-12 - "But as it is written, Eye hath not seen, nor ear heard, neither have entered into the heart of man, the things which God hath prepared for them that love him. But God hath revealed them unto us by his Spirit: for the Spirit searcheth all things, yea, the deep things of God. For what man knoweth the things of a man, save the spirit of man which is in him? even so the things of God knoweth no man, but the Spirit of God. Now we have received, not the spirit of the world, but the spirit which is of God; that we might know the things that are freely given to us of God."

VII. THE THREE-FOLD NATURE OF GOD – PART I

The composition of light

Let's briefly step back and summarize what we have learned up to this point. So far, we have observed quite an impressive list of properties regarding the physics of light that perfectly line up with the attributes of the God revealed in the Bible.

Throughout His Word, God relates Himself to light. We normally interpret these scriptures as completely spiritual in nature, meaning that we ascribe this connection to light to be nothing more than God relating to us that He illuminates spiritual truths, much like physical light illuminates the physical world around us. It is the premise of this book that the relationship between God and light is much deeper, as evidenced by a series of improbable parallels between the physics of light and other more profound and defining characteristics of God. It's almost as if God went out of his way to design the physics of light such that light would harbor a secret message that would one day be discovered when physicists probed its hidden depths. I have come to believe that the purpose of this message is to disclose the true identity and nature of the Creator.

So far, we have witnessed the following parallels:

God is Absolute ←→ Light is Absolute

God is Timeless ←→ Light is Timeless

God is Omnipotent ←→ Infinite Power is Required to Reach the Speed of Light

God is Omnipresent ←→ Light is Omnipresent

God Is Omniscient ←→ Light Can Transmit an Infinite Amount of Information

God's Mind is Concealed ←→ Light's Communication is Secret and Unbreakable

Now, we will see how the physics of light reveals to us the nature of God's specific and unique identify. Out of all of the gods mentioned in the different cultures of the world, which God is the real God? Did God leave a message in light that would reveal His true name?

Out of all of the properties of light, this may be the most telling in terms of revealing the unique identity of God. Without this property, one may still be left hanging with respect to discovering the identity of the true God. It's wonderful to know that God is absolute, timeless, infinite in power (omnipotent), omnipresent, and omniscient, but it's an entirely different matter to know His name.

To see this, we must now look at what light is composed of. Physicists debated over the course of the 20th century about the composition of light. One camp stated that light was a particle. Another camp stated that light was a wave. These two aspects of light may seem to be mutually exclusive and contradictory, and to a certain extent, they are. Particles and waves act much differently. Particles are discreet, tangible packets of matter/energy with a defined location and volume. Waves are not as discreetly defined. In a certain sense, waves are intangible. Waves diffusely spread out, diffract, refract, and add together (interference) in ways that particles do not. As more experiments were conducted on light, physicists concluded that light is both a wave and a particle. This property is called the wave-particle duality of light. The major differences between waves and particles are illustrated below:

Wave-Particle Duality of Light

Wave
- Spread Out
- Not a "tangible" object
- Exhibits interference, diffraction, and refraction

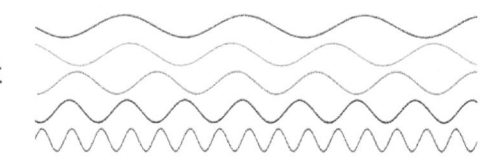

Particle – Photon
- Defined position and volume
- "Tangible" in nature
- Exhibits "collisions"

As scientists and theoreticians probed deeper, they found that the wave was actually divided into two different energy "fields". Scientists discovered that the wave nature of light stems from an oscillating electric field and an oscillating magnetic wave. For this reason, light is referred to as an "electromagnetic" wave. To complete the picture, the particle-like nature of light was attributed to a quantized packet of energy called a "photon". The photon is meant to carry the particle like nature of light.

More specifically, light is electromagnetic energy and it is composed of three components:

1) An oscillating electric wave that moves at the speed of light.
2) An oscillating magnetic wave that propagates at the speed of light in perfect synchrony and at right angles to the electric wave; and,
3) A "particle" like character termed a 'photon'.

You can see this three-fold structure of light illustrated in the figure below:

Three-Fold Nature of Light and God

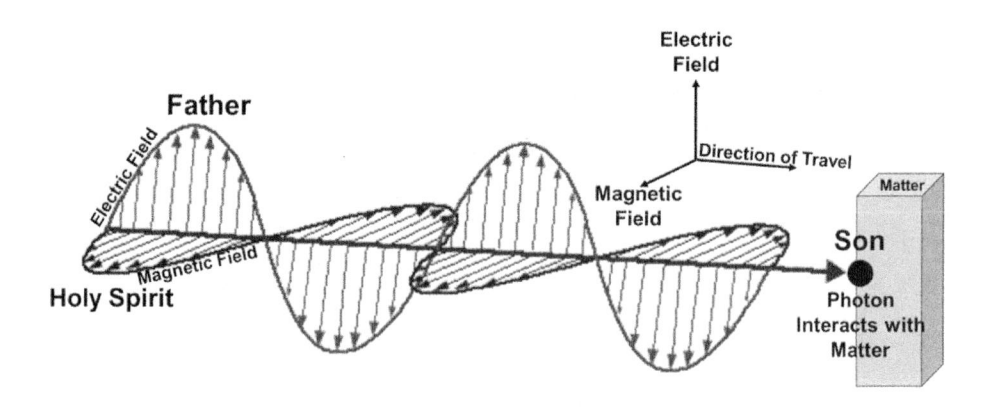

You can draw a direct parallel between each one of these components of light to the different manifestations of God as revealed in the Bible. You can assign the electric field to the Father, the magnetic field to the Holy Spirit, and the photon to the Son. I will describe these parallels in further detail throughout this chapter.

Electric Field ←→ Father
Magnetic Field ←→ Holy Spirit
Photon ←→ Son

Conversely, you could just as easily assign the magnetic field to the Father and the electric field to the Holy Spirit. If I had to choose between the two options, I would assign the Father to the electric field, and the Holy Spirit to the magnetic field. The reason for this choice has to do with the nature of electric fields and magnetic fields. Magnetic fields are always generated by a charge/electric field that is in motion. This is how an electromagnet works. As electrons move through the wire, their motion produces a magnetic field. This is also how permanent magnets work. As electrons whiz around in the atoms that make up the metal, they produce a magnetic field. We know that when God moves, He usually does this through His Holy Spirit. The Holy Spirit is how God moves in and through His people. This parallel leads me to assign the Father to the electric field, and the Holy Spirit to the magnetic field, but I certainly wouldn't be rigid about this particular distinction.

It is also worth noting that light is a self-perpetuating system. This means that after a light beam is created, no other additional energy is required to make the light beam continue through space forever! Why? Well, it has to do with Maxwell's equations. Essentially, these four equations describe the following: A wiggling electric field induces a wiggling magnetic field. The wiggling magnetic field, in turn, induces a wiggling electric field. And the wiggling electric field.... well, you get the picture. This goes on forever. The electric field and the magnetic field do this in perfect harmony such that each field perpetuates/generates the other field, and this cycle carries on ad infinitum. They both travel in the same direction forever, locked into the same frequency and phase through space at the highest speed known to man, 186,282 miles per second. Maxwell's equations describe how light self-perpetuates or self-sustains itself in this cyclic fashion. Once a light beam is let loose in the vacuum of space, it will travel forever with no additional energy input. In this sense, electromagnetic waves are infinite in nature. Just like light is self-sustaining, God is self-sustaining as He requires no additional or outside energy source to exist. God persists forever without change.

God is Self-Sustaining ←→ Light is Self-Sustaining

Building off of this, the wiggling electric field and wiggling magnetic field travel together in perfect synchronization and unison. Due to their self-sustaining interconnectedness, they travel together in lockstep, with the same frequency, wavelength, phase, and speed (as illustrated in the figure above). Just like light, we know that God is self-consistent, harmonious, congruent, and perfectly unified.

God is Unified ←→ Light is Unified

Now, the electric field and magnetic field are abstract expressions in another dimension for they are not visible or tangible. The electric field and magnetic field of light are invisible, intangible, ethereal, abstract expressions. As opposed to existing in a physically tangible format, the electric field and magnetic field are more like the nature of a mathematical equation. The electric field and magnetic field essentially live in a virtual space, but their effects are measured in our physical reality when light interacts with matter (more on this below). Since the scriptures tell us that the Father has never been seen at any time, and the Holy Spirit is an invisible Spirit, we can easily see how the Father and the Holy Spirit map to the electric field and magnetic field of light. They are invisible and intangible.

Now, it's only when the electric field and magnetic field interact with physical matter in some way that they become known to us. This interaction is mediated by the third component, the photon. The photon carries the particle-like nature of light. The photon is a manifestation of light into our visible reality that allows us to experience light. Einstein won a Noble prize in 1921 for his description of the photoelectric effect, which was a direct demonstration that light is divided into indivisible subunits, or discreet packets, called photons. The photon physically interacts with matter around us and directly interacts with the cells in the retina of our eye. In a conceptual sense, it is proper to think of a photon as a tiny particle. The photon is a visible and "tangible" manifestation of something that is invisible.

From the scriptures, we know that Jesus is the Son of God, and He is the visible manifestation of the invisible God. Jesus is the Word become flesh. GOD WITH US. A God we can touch, feel, and SEE. Jesus is the part of God that is seen by the physical universe. The invisible God has become visible. God has now interacted with man, just like an invisible electromagnetic field can only interact with us via a photon.

As mentioned, there is a great mystery that surrounds the nature of light that scientists call the "wave-particle" duality of light. In one experiment, light appears to behave like a wave with no particle-like characteristics. If you measure light in another way, it appears to behave like a particle with no wave-like characteristics. So, scientists say that light is somehow both a wave (the electric and magnetic field), and a particle (the photon) at the same time. This paradoxical wave-particle duality of light directly parallels the same mystery regarding the God-man duality of God. How can God be both fully God and fully man at the same time? How can light be both an invisible wave and a physically tangible discrete particle (photon) at the same time?

In the same way that physicists wrestle with the paradoxical wave-particle duality of light, theologians wrestle with the paradoxical God-man duality of God, where Jesus is understood to be fully God and fully man. This will be discussed in more depth in a subsequent chapter.

The key to the full nature of light is that all three components of light require one another. It is impossible to isolate one aspect of light without the other two somehow being present. A waving electric field has never been found by itself. This is because a waving electric field always generates a waving magnetic field. Similarly, a waving magnetic field has ever been found by itself. This is because a waving magnetic field always generates a waving electric field. When there is a waving electric and magnetic field, it naturally comes along with the photon when it interacts with our physical universe (matter).

To complete the picture, a photon is always present with an electromagnetic field. In fact, the photon is said to the be the "mediator" or force carrier of the electromagnetic force! Similarly, Jesus is said to be the "mediator" between God and man:

> *I Timothy 2:5* – *"For there is one God, and one mediator between God and men, the man Christ Jesus;"*

If you have any one of the three, then you find the other two. It is impossible to isolate just one component. Even though all three components depend on one another, each component is fundamentally different. A magnetic field is fundamentally different from an electric field, and both are fundamentally different from a photon. Even though they are different, you can never find them alone.

I submit that this is a perfect picture of the Trinity: Father, Son, and Holy Spirit. Each member of the Trinity is a unique manifestation of God with their own characteristics; yet they are unanimously one and mysteriously intertwined in an inextricable way. This three-fold picture of God and light is illustrated below:

Three-Fold Nature of God and Light

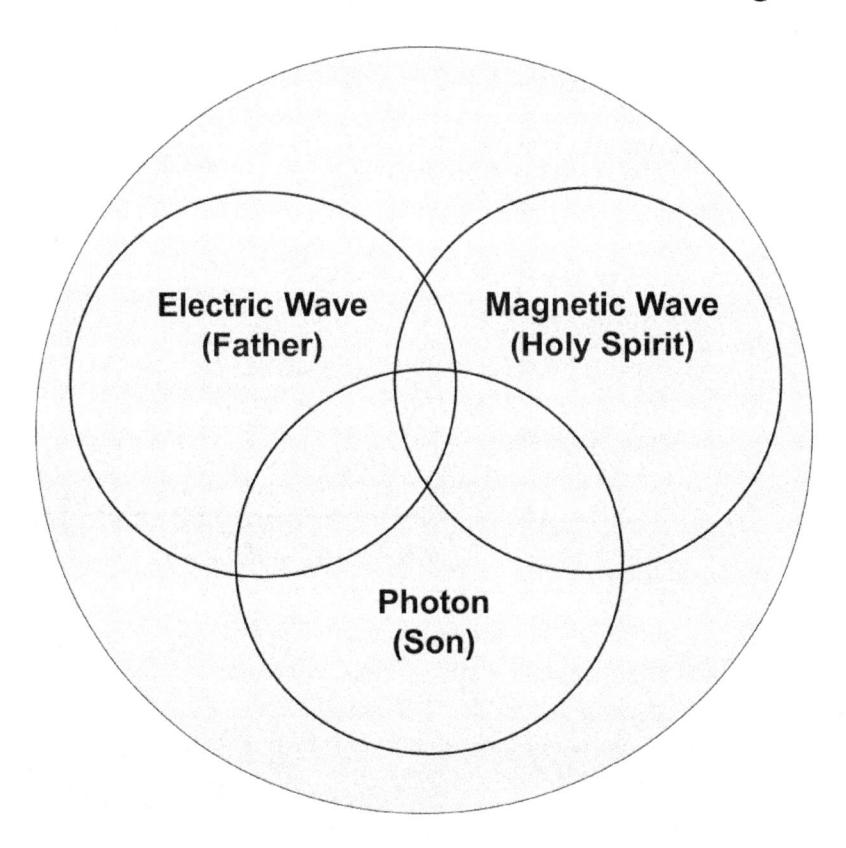

- Electric Wave (Father)
- Magnetic Wave (Holy Spirit)
- Photon (Son)

The Father, Son, and Holy Spirit are all part of one nature such that you can never find them separate. This is why Jesus claimed that He and the Father are one, and that if you have seen Him, then you have seen the Father. In a beautiful way, the physics of light is a physical metaphor that parallels the three-fold identify of the God described in the Bible.

God is a Trinity ←→ Light is a Trinity (In Composition)

Father (Invisible, Intangible) ←→ Electric Field (Invisible, Intangible)
Holy Spirit (Invisible, Intangible) ←→ Magnetic Field (Invisible, Intangible)
Jesus (Visible, Tangible) ←→ Photon (Visible, Tangible)

> Genesis 1:26 – "Then God said, "Let us make man in **our** image, after **our** likeness."
>
> John 1: 1-5, 9, 10, 12, 14 - "In the beginning was the Word, and the Word was with God, and the Word was God. The same was in the beginning with God. All things were made by him; and without him was not any thing made that was made. In him was life; and the life was the light of men. And the light shineth in darkness; and the darkness comprehended it not.......That was the true Light, which lighteth every man that cometh into the world. He was in the world, and the world was made by him, and the world knew him not.... But as many as received him, to them gave he power to become the sons of God, even to them that believe on his name.... And the Word was made flesh, and dwelt among us, (and we beheld his glory, the glory as of the only begotten of the Father,) full of grace and truth."

John 1: 18 - "No man hath seen God at any time, the only begotten Son, which is in the bosom of the Father, he hath declared him."

John 10:30 - "I and my Father are one."

Matthew 28:19 - "Go therefore and make disciples of all nations, baptizing them in the name of the Father and of the Son and of the Holy Spirit."

1 Corinthians 8:6 - "yet for us there is one God, the Father, from whom are all things and for whom we exist, and one Lord, Jesus Christ, through whom are all things and through whom we exist."

John 14: 16-17 - "And I will pray the Father, and he shall give you another Comforter, that he may abide with you for ever; Even the Spirit of truth; whom the world cannot receive, because it seeth him not, neither knoweth him: but ye know him; for he dwelleth with you, and shall be in you."

Colossians 2:9 – "For in him (Christ) dwelleth all the fulness of the Godhead bodily."

Isaiah 9:6 – "For unto us a child is born, unto us a son is given: and the government shall be upon his shoulder: and his name shall be called Wonderful, Counsellor, The mighty God, The everlasting Father, The Prince of Peace."

Philippians 2:5-8 "Who, being in very nature God, did not consider equality with God something to be used to his own advantage; rather, he made himself nothing by taking the very nature of a servant, being made in human likeness. And being found in appearance as a man, he humbled himself by becoming obedient to death—even death on a cross!"

> *I Timothy 2:5 – "For there is one God, and one mediator between God and men, the man Christ Jesus;"*
>
> *Colossians 1:14-17 –"In whom (Christ) we have redemption through his blood, even the forgiveness of sins: Who is the image of the invisible God, the firstborn of every creature: For by him (Christ) were all things created, that are in heaven, and that are in earth, visible and invisible, whether they be thrones, or dominions, or principalities, or powers: all things were created by him, and for him: And he is before all things, and by him all things consist."*

(As a side note about the Trinity: I realize that there are people who reject the concept of the Trinity. First, this word is not used throughout the scripture. This term was introduced into Christianity after the scriptures were recorded. However, this word "Trinity" is just a placeholder or pseudonym that we use to communicate a truth that is revealed throughout the scripture in many texts; namely, that God is three-fold in nature: Father, Son, and Holy Spirit. Another objection to the Trinity is the fact that Jesus claimed that the Father was greater than Himself. In response to this, it is necessary to place those statements in context with the other scriptures where Jesus claims that if you have seen Him, then you have seen the Father, and that He and the Father are one. In fact, when Jesus made these claims, the scripture tells us that the Pharisees understood that He was making Himself equal with God the Father, and this claim of equality was the premise that was used by them to condemn Him to crucifixion. In other parts of scripture, Jesus forgives the sins of others: again, the Pharisees recognized that this was a statement of His Divinity, as only God can forgive sins. To understand the mystery of the three-fold nature of God, ALL scripture has to be taken together, not just isolated statements. As examples, please read the scriptures above: they all point to the three-fold nature of God. There is a reason Jesus claims to be less than God in one breath, and then seemingly contradicts Himself by claiming (and acting) as if He were equal to God in another. The reason is because Jesus was both fully God and fully man simultaneously. He was finite and infinite all rolled into one. This is a paradox, and true paradoxes always generate seemingly contradictory statements, but there is no real contradiction. So, when Jesus claimed that He was less than God, He was referring to his humanity. When He claimed He was equal to the Father, He was referring to His Divinity. These statements seem to be contradictions, but they are not contradictions. There is a difference between a true contradiction, and a paradox. This paradox of how Jesus was fully God and fully man is the same paradox that is presented with the wave-particle duality of light, which is handled in more detail in a different chapter).

VIII. THE THREE-FOLD NATURE OF GOD – PART II

The color of light

In the last chapter, we saw how there is a direct correlation between the three-fold nature of God as described in the Bible and the composition of light. We will now add to this topic by showing that the Trinity of God is also expressed in the color expression of light. I would propose that God really wanted to drive this point home, because it is this three-fold property of light that authenticates His true identity.

There are literally an infinite number of colors along the color spectrum. Any one of these infinite colors can be made by mixing or adding together precisely three colors. In additive color mixing, three primary colors are combined in different ratios to create an infinite spectrum of colors. Additive color mixing was first formulated by the Young–Helmholtz theory of trichromatic color vision, but James Clerk Maxwell is usually given credit for this field because of his experimental work.

It is important to note that the perception of color is something that is categorically defined by our brain. This is important because it uniquely connects this physical law of color addition to the structures of human biology. Also, along this line, our retina contains three different color photoreceptors, each which respond to a different wavelength/color of light. This trinity of photoreceptors corresponds with the trinity of colors that add together to give all other possible colors along the color spectrum.

In additive color mixing, any color can be produced by combining some ratio of the three primary colors: red, green, and blue. If you add all three primary colors together in equal combination, the result is pure white light. Millions upon millions of different colors and shades of colors can be produced by combining just these three colors. This is illustrated below:

Three-Fold Color Spectrum

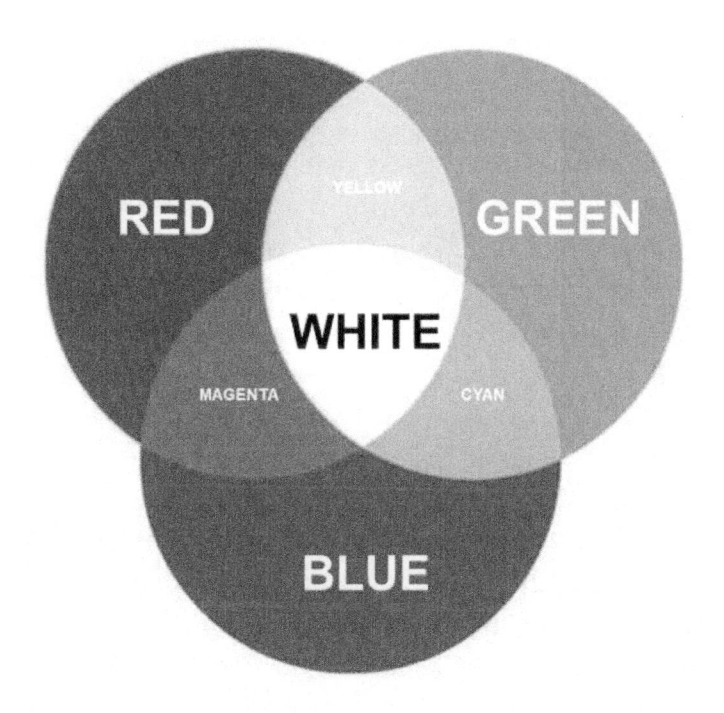

Relating these physical laws to the spiritual nature of God, a very exciting picture emerges; namely, that this three-fold God, who is Spiritual Light, illuminates the world around us in an infinite number of spiritual colors of revelation.

When these three colors are perfectly combined equally, pure white light emerges. It is also important to understand that to observe the true color of an object, it is necessary to shine pure white light on the object. If you shine yellow light on a blue ball, for example, it will not appear blue. White light reveals and exposes everything for what it truly is. In the same way, God, who is a perfect combination of the members of the Trinity (Father, Son, and Holy Spirit), shines pure white Spiritual Light on any person and reveals the true nature of that person. Just like physical light illuminates the physical world around us, God illuminates the spiritual world.

Also, we learn that this connection between the spiritual nature of God and the physics of light is hardwired into our biology and brain chemistry. This indicates that this truth about light was specifically designed to communicate directly to us. In the previous chapter, we saw that one member of the Trinity, the Son, is the manifestation of God in the flesh. This Son, Jesus, was surely able to see these colors with His own photoreceptors, and experience white light the same way we do. The very fact that God wrapped Himself in human flesh to dwell with us tells us something: It demonstrates to us that we are a very special part of God's Creation, and that the physics running the universe was designed to communicate His nature to us. We are the centerpiece.

Let's pause for a moment and consider everything we have learned so far. It's quite amazing, isn't it? What are the chances that the physics of light would just happen to line up so perfectly with what we know to be true about the True God? Remember, these physical principles underlying the physics of light were not known thousands of years ago when the scriptures were written. These principles were not known until the modern era. Yet, the scriptures are profoundly replete with the peculiar connection between God and light. Again, we see that the universe was created with a physics that identifies the true nature of the Creator. Also, the message being communicated through this physics is that humans hold a very special place in this creation, as our biochemistry is designed to perceive the parts of God's nature that He chooses to reveal to us.

God is a Trinity ←→ Light is a Trinity (In Color)

God Illuminates the Spiritual World ←→ Light Illuminates the Physical World

> *Genesis 1:3-4 - "And God said, Let there be light: and there was light. And God saw the light, that it was good: and God divided the light from the darkness."*
>
> *Psalm 119:105 – "Thy word is a lamp unto my feet, and a light unto my path."*
>
> *Psalm 139: 11-12 – "If I say, Surely the darkness shall cover me; even the night shall be light about me. Yea, the darkness hideth not from thee; but the night shineth as the day: the darkness and the light are both alike to thee."*
>
> *John 9:5 – "As long as I (Christ) am in the world, I am the light of the world."*
>
> *John 1:9 - "That (Christ) was the true Light, which lighteth every man that cometh into the world."*

John 8:12 – "I (Christ) am the light of the world: he that followeth me shall not walk in darkness, but shall have the light of life."

I John 1:5-9 - This then is the message which we have heard of him, and declare unto you, that God is light, and in him is no darkness at all. If we say that we have fellowship with him, and walk in darkness, we lie, and do not the truth: But if we walk in the light, as he is in the light, we have fellowship one with another, and the blood of Jesus Christ his Son cleanseth us from all sin. If we say that we have no sin, we deceive ourselves, and the truth is not in us. If we confess our sins, he is faithful and just to forgive us our sins, and to cleanse us from all unrighteousness.

IX. THE THREE-FOLD NATURE OF GOD – PART III

The calibration of light

In the past couple of chapters, we looked at the three-fold pattern that is revealed in the physics of light. First, we observed this three-fold nature of light expressed in the composition of light, and we saw how each component of light lined up with a unique personality of the Trinity. Secondly, we witnessed another three-fold property of light expressed in the color space of light, and we related this to the spiritually illuminating power of the Triune God.

Now, we will continue that line of thought by looking at one of the most famous equations in physics:

$$E = mc^2$$

This equation, formulated by Einstein, accounts for the amount of energy that is present in a bit of matter. In fact, it states that matter and energy can be converted from one form to the other. This equation demonstrated that mass and energy are essentially equivalent. This equation shows that there is an incredible amount of energy in a small amount of matter. Why? In order to calculate energy, the speed of light (a very large number) is squared before it multiplies mass. This formula is what accounts for the incredible release of energy associated with splitting the atom. The physics underlying this formula is what drives the energy production of stars, and the energy released during the splitting of the atom to make nuclear energy possible. This formula also states that when you go jogging, your mass is increased to account for the extra kinetic energy of your body in motion. In this case, the amount of this mass increase associated with your increased kinetic energy is so small you couldn't measure it, but it is there nonetheless!

Now, look at the equation carefully. Do you notice the parameter that calibrates the amount of energy contained in a certain quantity of matter? That's right, it's the speed of light. Light is the standard that relates the interconversion between mass and energy.

In the graphic below, I illustrate how the Trinity can be mapped to this formula.

Mapping the Trinity onto $E=mc^2$

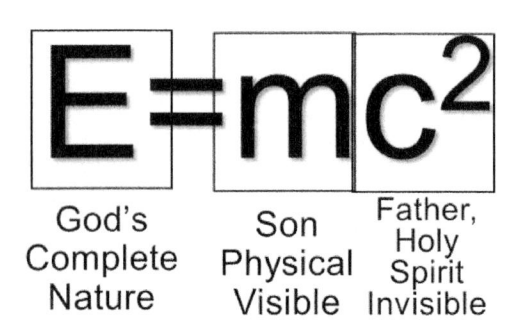

$$E=mc^2$$

| God's Complete Nature | Son Physical Visible | Father, Holy Spirit Invisible |

Energy is related to God's complete nature or essence. Energy in this equation embodies or symbolizes the totality of God. The equation reveals to us that His complete nature is made up of matter (mass), and the speed of light "c" times the speed of light "c".

Notice that mass is something that is physical and tangible. The Son, Jesus, is the physical, tangible manifestation of God. So, it makes sense to relate mass, m, to the Son. This is the physicality of God.

Now, "c" in this equation represents light. The electromagnetic wave the comprises light is not tangible. As I presented in a previous chapter, the electric wave and the magnetic wave are ethereal, intangible, invisible fields that travel through space at the speed of light. Notice that the speed of light is represented twice in the equation because it is squared. Therefore, we can relate the Father to the first c, and the Holy Spirit to the second c. The Father and the Holy Spirit are also intangible and invisible.

These 3 components of God's character (Father, Son, and Holy Spirit) make up the entire nature, "E", of God. Also, notice that there are **three** fundamental ideas accounted for in this formula: energy, mass, and light. Here we see, in very simple and elegant form, the Trinity of God represented in one of the most important equations of physics that defines the relationship between mass, energy, and light. And the speed of light, being an absolute value, calibrates how energy and mass are related.

In previous chapters that discussed the Special Theory of Relativity, we saw how light calibrates time, space, and matter. We now see that light also calibrates energy. Essentially, light is the absolute standard that calibrates everything in the universe.

Light Calibrates
Time
Space
Matter
Energy

Since matter (mass) and energy are equivalent, we can consolidate this to:

Light Calibrates
Time
Space
Matter/Energy

Therefore, we can now see that light, which is a Trinity, calibrates ALL of the fundamental building blocks of the universe, which are, themselves, a Trinity (Time, Space, Matter/Energy). Since God is using physical light as a metaphor for His nature, we see that God, like light, is a Trinity. In addition, we see that God, like light, is the absolute standard that calibrates and defines ALL of creation. God is Absolute Truth. Everything good, including Righteousness, Holiness, Perfection, and Purity, are all defined by Him.

This amazing parallel is illustrated below:

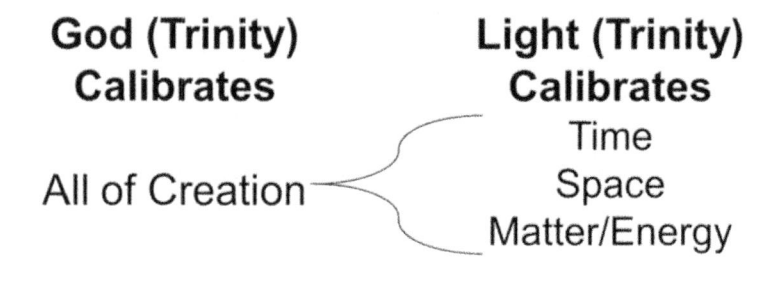

God (Trinity) Calibrates

All of Creation

Light (Trinity) Calibrates
Time
Space
Matter/Energy

God Calibrates All Creation ←→ Light Calibrates Time, Space, and Matter/Energy

> *John 14:6 - "Jesus saith unto him, I am the way, the truth, and the life: no man cometh unto the Father, but by me."*
>
> *Romans 3:4 – "let God be true, but every man a liar;"*
>
> *Psalm 19:7-10 – "The law of the LORD is perfect, converting the soul: the testimony of the LORD is sure, making wise the simple. The statutes of the LORD are right, rejoicing the heart: the commandment of the LORD is pure, enlightening the eyes. The fear of the LORD is clean, enduring for ever: the judgments of the LORD are true and righteous altogether. More to be desired are they than gold, yea, than much fine gold: sweeter also than honey and the honeycomb."*
>
> *Matthew 24:35 – "Heaven and earth shall pass away, but my words shall not pass away."*

X. THE THREE-FOLD NATURE OF GOD -PART IV

The spectrum of light

Now we will take a look at the different frequencies of light. As described in previous chapters, light is an electromagnetic wave: i.e., an electric field that is copropagating with a magnetic field through space at the speed of light.

All waves have a frequency and a wavelength. The frequency is a measure of the number of times that the field vibrates back and forth in a certain unit of time (the unit of measurement is Hertz). For example, if a wave vibrates up and down 50,000 times in one second, then its frequency is 50,000 Hertz. The wavelength is the distance in space between the peaks of the wave, or between the valleys of the wave, and is measured in meters. Frequency and wavelength are inversely related. If the frequency is high, then the distance between peaks is small, and if the frequency is low, then the distance between peaks is large.

The electromagnetic spectrum is a scale that measures the frequency of light and categorizes light based on that frequency. This is shown in the graphic below:

Electromagnetic Spectrum

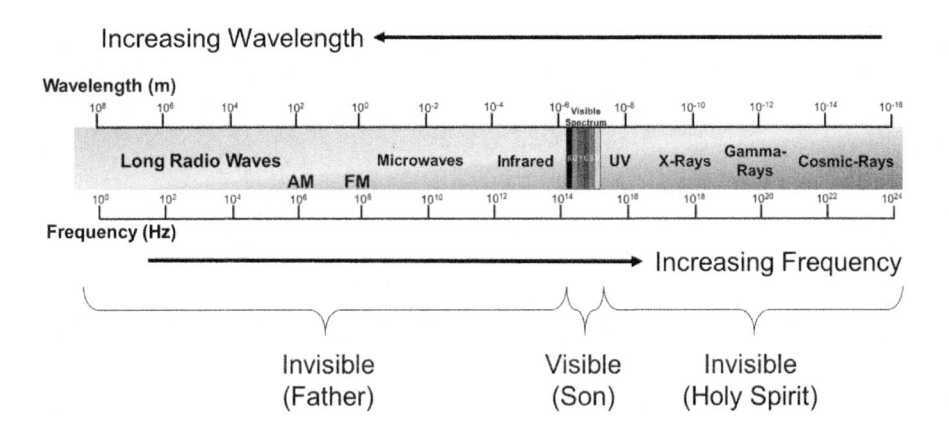

Wavelength (m): 10^8, 10^6, 10^4, 10^2, 10^0, 10^{-2}, 10^{-4}, 10^{-6} Visible Spectrum, 10^{-8}, 10^{-10}, 10^{-12}, 10^{-14}, 10^{-16}

Increasing Wavelength ←

Long Radio Waves | AM | FM | Microwaves | Infrared | UV | X-Rays | Gamma-Rays | Cosmic-Rays

Frequency (Hz): 10^0, 10^2, 10^4, 10^6, 10^8, 10^{10}, 10^{12}, 10^{14}, 10^{16}, 10^{18}, 10^{20}, 10^{22}, 10^{24}

→ Increasing Frequency

Invisible (Father) | Visible (Son) | Invisible (Holy Spirit)

At very low frequencies, light is a radio wave. This band of lower frequencies is used by radios. As you move up in frequency, you find microwaves (microwave ovens and cell phone communication use this band of frequencies), and then slightly higher you find infrared waves (heat energy). Directly adjacent to infrared is the optical spectrum. At this frequency, the photoreceptors in our retina are capable of sensing these electromagnetic waves. In the visible spectrum, as the frequency increases, the colors change sequentially from red, to orange, to yellow, to green, to blue, to indigo, to violet. The ultraviolet range occurs at slightly higher frequencies than the visible spectrum. Above UV, you find the realm of X-rays, gamma rays, and finally cosmic rays.

As you can see, the range of frequencies that humans can actually see fall within a very narrow window or thin sliver of the entire electromagnetic spectrum. The overwhelming majority of the electromagnetic spectrum is completely invisible! But it's all made of the same stuff. There is nothing fundamentally different about the nature of the light at different frequencies – it's all the same: electromagnetic energy. There's virtually an unlimited number of frequencies above and below the optical spectrum. The spectrum is, for all practical purposes, infinite. But even within the narrow slice of the visible spectrum, there are an infinite number of frequencies and corresponding colors! We saw in a previous chapter how an infinite number of colors in this visible part of the spectrum can be created by the combination of only three colors.

Relating this to God, we know from the scriptures that God is infinite in knowledge, wisdom, power, righteousness, and love. But God only chooses to reveal to us a very narrow sliver of Himself (the optical spectrum). Christ is "visible" image of the "invisible" God, but Christ possesses the entire fullness of God. The Father, the Holy Spirit, and the Son are all part of same Person, just like all parts of the electromagnetic spectrum are made from the same electromagnetic energy.

Again, we see the Trinity being reflected in the physics of light, because we can essentially divide the electromagnetic spectrum into 3 regimes: Frequencies below the optical spectrum (Invisible); Frequencies within the optical spectrum (Visible); and, Frequencies above the optical spectrum (Invisible). Two of the components are invisible, and one is visible, just like the Trinity. To represent this, I have mapped each member of the Trinity onto the electromagnetic spectrum above.

At any given time, we can only see a very limited sliver of God and what He is doing. Even though we cannot see Him, He is there. He is there in all of His infinite glory.

There are a number of parallels between God's nature and light that we can learn from the electromagnetic spectrum.

God is a Trinity ←→ Light is a Trinity (In Spectrum)

Father (Invisible) ←→ Frequencies Below Visible (Invisible)
Holy Spirit (Invisible) ←→ Frequencies Above Visible (Invisible)
Jesus (Visible) ←→ Frequencies Within the Visible (Visible)

God is Infinite ←→ Light Has Infinite **Frequencies**

God Reveals a Portion of Himself to Us ←→ Light Reveals a Portion of Itself to Us

> *Romans 1:19-20 - "...what may be known about God is plain to them, because God has made it plain to them. For since the creation of the world God's invisible qualities—his eternal power and divine nature—have been clearly seen, being understood from what has been made, so that people are without excuse."*
>
> *Colossians 1:14-17 – "In whom (Christ) we have redemption through his blood, even the forgiveness of sins: Who is the image of the invisible God, the firstborn of every creature: For by him (Christ) were all things created, that are in heaven, and that are in earth, visible and invisible, whether they be thrones, or dominions, or principalities, or powers: all things were created by him, and for him: And he is before all things, and by him all things consist."*

XI. HOLDING ALL THINGS TOGETHER

The reason we don't fly apart

We are studying how the physics of light communicates a complete story about God's nature and identity. God continually refers to Himself as "light" throughout the scripture, and we are finding that this declaration is much deeper and expansive than the typical interpretation of this truth. We are discovering that the unique identity of God is authenticated in a code that exists in the physics of light. God left behind a fingerprint that paints a very clear picture about His character and identity, and this authentication is seen in multiple layers. Now, as we learn more about light, the story becomes even more interesting.

As discussed in previous chapters, light is electromagnetic energy. Light is composed of an electric field and a magnetic field which oscillate together in perfect unity and move through space at the speed of light, c.

Let us turn our attention towards matter. Matter consists of individual atoms which are arranged in a specific configuration. A combination of atoms that are bound together into a larger unit is called a molecule. Do you know what holds all atoms and molecules together?

The answer is light!

An atom consists of a combination of electrons, protons, and neutrons. Protons and neutrons comprise the central nucleus of the atom, while the electrons whiz around the nucleus in a cloud. The electromagnetic force binds electrons to the nucleus of the atom (the nucleus is held together by a different force, but that is a story for another day).

In addition to holding individual atoms together, the electromagnetic force is responsible for the binding energy that connects atoms together to form molecules and solids (via covalent bonds, ionic bonds, and metallic bonds). Also, the electromagnetic force is responsible for the interactive forces between molecules (via ion-dipole, hydrogen bonding, and van der Waals forces). If you recall, the photon is the mediator or the force carrier of the electromagnetic force, and it is the electromagnetic force that forms the basis of molecular bonds.

In other words, all matter is being held together by light energy. Every single atom depends on light energy to bind it together; every single molecule depends on light energy to hold the atoms together; and every single molecule depends on light energy to bind molecules to one another.

This includes every single molecule, every single system of molecules, every atom and molecule in our body, every atom and molecule that makes up the earth, and every atom and molecule throughout the universe. That's pretty much everything. All matter is being held together by light. Light is the glue that holds all things together. This is illustrated below:

Light Energy Holds Atoms and Molecules Together

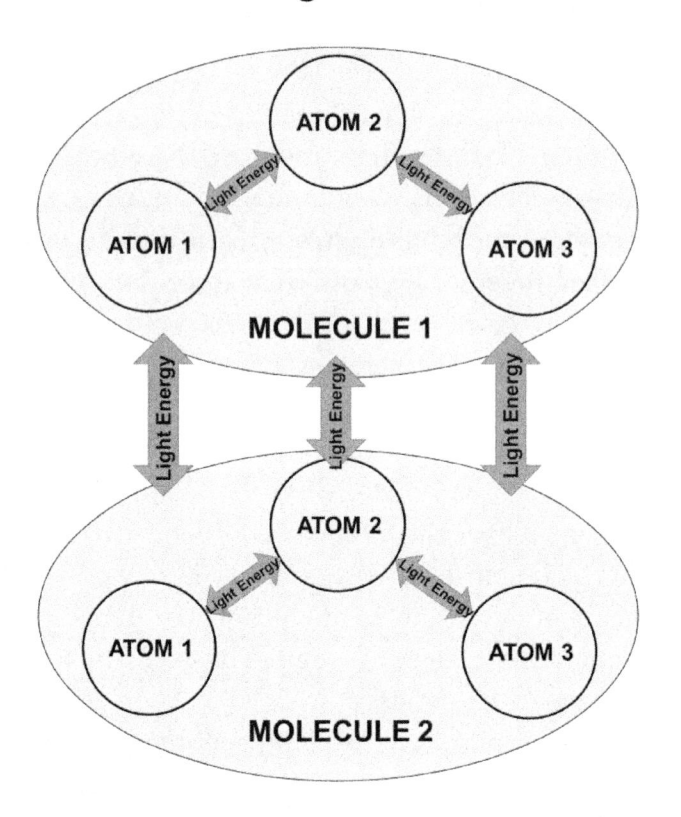

And this physical truth about light is a perfect metaphor for what God performs both physically and spiritually on a continual basis. God holds the entire universe together, and He holds our entire life together, even when we feel like it is spinning out of control.

God Holds All Things Together ←→ Light Holds All Things Together

> *Colossians 1:17 - "He is before all things, and in him all things hold together."*
>
> *Hebrews 1:3 – "Who being the brightness of his glory, and the express image of his person, and upholding all things by the word of his power, when he had by himself purged our sins, sat down on the right hand of the Majesty on high:"*

XII. GOD BECAME MAN

An infinitely finite and finitely infinite paradox

As introduced earlier, there is one very peculiar property about light that reveals to us something very profound about the unique identity of God.

During the 20th century, exceptionally intelligent theoretical physicists wrestled with whether light was made up tiny particles or waves. This was not a new debate. Democritus from the 5th century BC viewed light as a combination of indivisible sub-components (i.e., particles), while René Descartes from the 17th century modeled light as a wave. Sir Isaac Newton viewed light as particles which he developed in his corpuscular theory of light. Around the same time of Newton, Robert Hooke, Christiaan Huygens and Augustin-Jean Fresnel all championed the wave-model of light. In 1801, Thomas Young's famous double-slit experiment demonstrated that light could interfere with itself, a property only exhibited by waves. In 1873, James Clerk Maxwell formulated what became known as Maxwell's equations, which described the wave nature of light in terms of co-propagating waves of oscillating electric and magnetic fields. Then, at the turn of the 20th century, Max Planck demonstrated that the electromagnetic field can be discreetly quantized into integer packets of energy. Einstein's description of the photoelectric effect in 1905, for which he won the Nobel Prize in 1921, was based on the discreet particle-like nature of light. These packets of energy ultimately became known as the photon, and it is the photon which carries the particle-like nature of light.

On the surface, these two models that describe the essence of light are contradictory. This paradoxical difference between the two models led to much healthy debate. To this day, there is no definitive winner, as both models are true.

On one hand, particles are discrete, defined in space and volume, countable, and tangible. On the other hand, waves are spread out, continuously changing, and intangible.

Particles transfer momentum to one another when they collide. Two particles do not occupy the same volume of space at the same time. On the other hand, waves pass through one another, interfere, diffract, and refract. Waves can occupy the same volume of space at the same time.

The interference of light is a unique property of waves that is counterintuitive to our understanding of tangible particles. When multiple waves are combined together, the net result is determined by the algebraic addition of the waves at every point in space. In other words, if at every point in space wave A is exactly equal in intensity but opposite in sign to wave B, the sum of these two waves will be zero. This is called destructive interference. Fundamentally, this means that two light beams could add up to zero (i.e., blackness)! Conversely, if the two waves are in phase such that their sum is greater than either wave, this is called constructive interference. This is illustrated in the figure below:

Constructive Interference

Destructive Interference

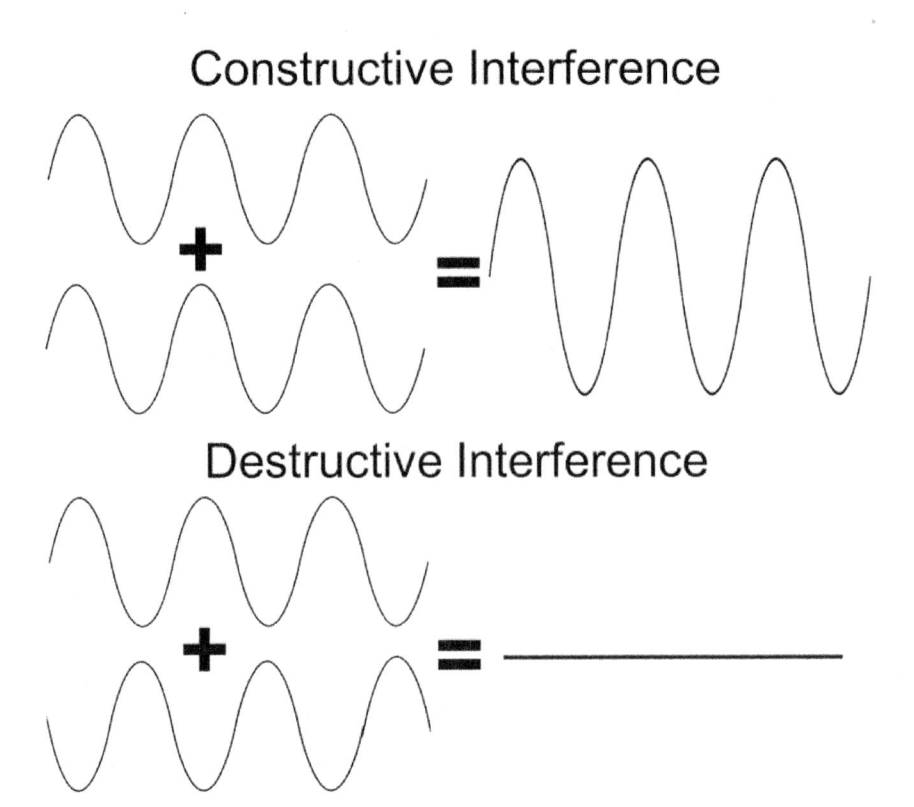

Based on these incompatible differences between waves and particles, it would seemingly be impossible for something to be both a particle and a wave at the same time. When physicists performed experiments on light, the behavior of light could manifest as a wave or as a particle depending on the nature of the experiment. Scientists were perplexed because this was thought to be a paradox. This infamous property of light became known as the wave-particle duality of light. In fact, it was later shown that all matter has this same duality. Regarding this duality, Albert Einstein wrote:

> *"It seems as though we must use sometimes the one theory and sometimes the other, while at times we may use either. We are faced with a new kind of difficulty. We have two contradictory pictures of reality; separately neither of them fully explains the phenomena of light, but together they do."* – Albert Einstein

To this day, physicists have not resolved this paradox but have learned to live with its ramifications.

If the characteristics of God are conveyed in the physics of light, what can we learn about God from this duality? Remember, God repeatedly referred to Himself as light in the scripture, and it is reasonable to ask if there are principles that govern the physics of light that correspond directly with specific aspects of God's nature. There are many paradoxical aspects of God's nature, but probably the most prominent dichotomy is the one presented by the claim that God is fully God and fully man.

According to the scriptures, we know that God became flesh in the person of Jesus Christ, His Son. When God took on flesh, He became tangible. The scriptures claim that in Him (Christ) dwells all the fullness of the Godhead bodily (Colossians 2:9). Jesus Christ, a finite man, has all of the infinity of God dwelling in Him. He is fully God, and fully man. He is not half God, and half man.

This creates a paradox because the finite and the infinite are dwelling together simultaneously. This is the same paradox which light presents us.

The photon is the particle like manifestation of light. The photon is "tangible", finite, and discrete, and these properties correspond to the Son. In contrast to the particle nature of light, the wavelike character of light refers to the electric wave and the magnetic wave, which co-exist, and self-generate one another in perfect unity through empty space. These co-propagating waves travel through space at the speed of light and will continue to travel through space. These waves are "invisible' and "intangible". As we discussed in an earlier chapter, these waves perfectly correspond with the Father and the Holy Spirit, who are both invisible, intangible, and infinite.

Light is fully wave, and fully particle. God is fully God, and fully man. The infinite coexists with the finite, and the finite coexists with the infinite in a paradoxical relationship.

Jesus is Fully God and Fully Man ←→ Light is Fully Wave and Fully Particle

> *John 1:14 – "And the Word became flesh and dwelt among us, and we have seen his glory, glory as of the only Son from the Father, full of grace and truth."*
>
> *John 10:30 – "I and my Father are one."*
>
> *Colossians 1:9 – "For in him dwelleth all the fulness of the Godhead bodily."*

XIII. BIRTH, DEATH, AND RESURRECTION

The defining moment in history

So far, we have seen how the physics of light directly corresponds with the characteristics associated with the God revealed in the Bible. The perfect alignment of the strikingly peculiar properties of light to the strikingly peculiar characteristics of God is quite extraordinary.

The Bible and other historical records that capture the life of Christ record a remarkable story that is unlike any other. In these records, God visits His creation by wrapping Himself in flesh through the physical birth of Jesus Christ on the earth. As God's one and only Son, Jesus lived on the earth approximately two-thousand years ago and interacted with thousands of different people. Central to this story is his death by crucifixion and His subsequent resurrection from the dead three days after His death. Hundreds of different people witnessed His death and resurrection, and many of these people were so radically transformed by this experience that it compelled them to tell the world about what they witnessed, even when they knew that they would be persecuted and probably die in the process.

It can be reasoned that this is probably the most distinctive or identifying feature of the Christian God, and given all of the other parallels observed so far, it is reasonable to assume that the birth, death, and resurrection of Jesus Christ should be recorded in in the physics of light. But how? .... One day, it hit me like a ton of bricks.

When light interacts with physical matter, it interacts through the photon-nature of light. This was discussed in previous chapters. If you recall, the photon, being a tangible manifestation of an invisible wave, directly corresponds to Jesus, who is God in the flesh. When invisible electromagnetic energy interacts with physical matter, this interaction is mediated by the photon. How does this interaction take place?

To understand how this interaction takes place, we need to look at the building block of matter, the atom. An atom is comprised of a nucleus (which consists of protons and neutrons), and a cloud of electrons that whiz around the nucleus. When a photon from a beam of light illuminates a piece of matter, the photon interacts with the electron cloud that is on the outside of the atom. If the photon has the right frequency, it will be completely absorbed by the electron, and the energy from the photon will cause the electron to jump to a higher energy level. After the electron jumps to the higher energy level, the light beam (electromagnetic wave) is nowhere to be found because it was essentially "absorbed" by the electron. This can be seen in the illustration below. With me so far?

Light Interacting with Matter

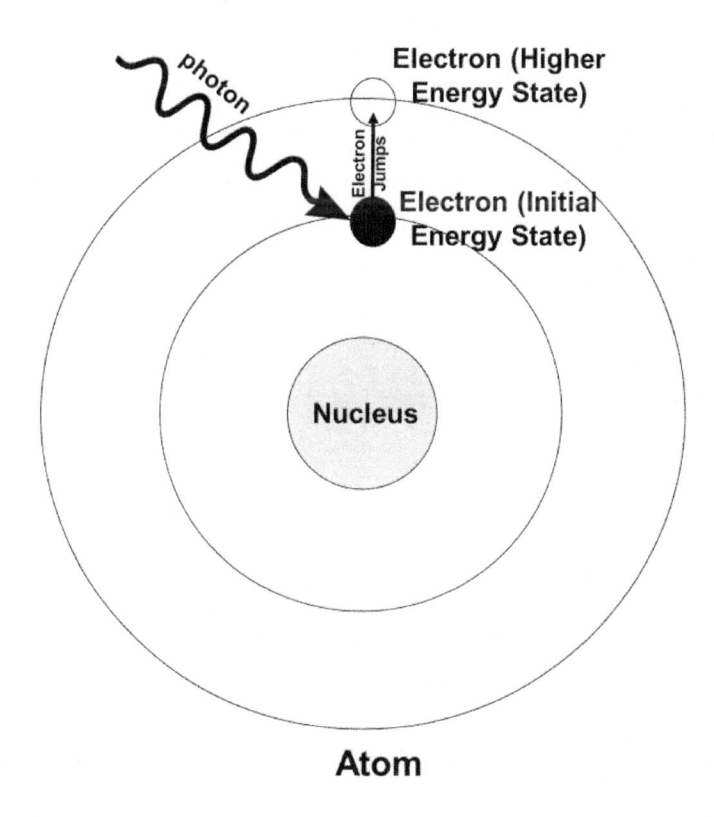

Electron (Higher Energy State)
photon
Electron Jumps
Electron (Initial Energy State)
Nucleus

Atom

Now, at some point in the future this electron can "decide" to jump back down to the lower energy level. By doing so, it will release a photon of the same energy and frequency of the original photon that was absorbed. When this occurs, a light beam is released that travels through space as an electromagnetic wave. This is illustrated below:

Matter Releasing Light

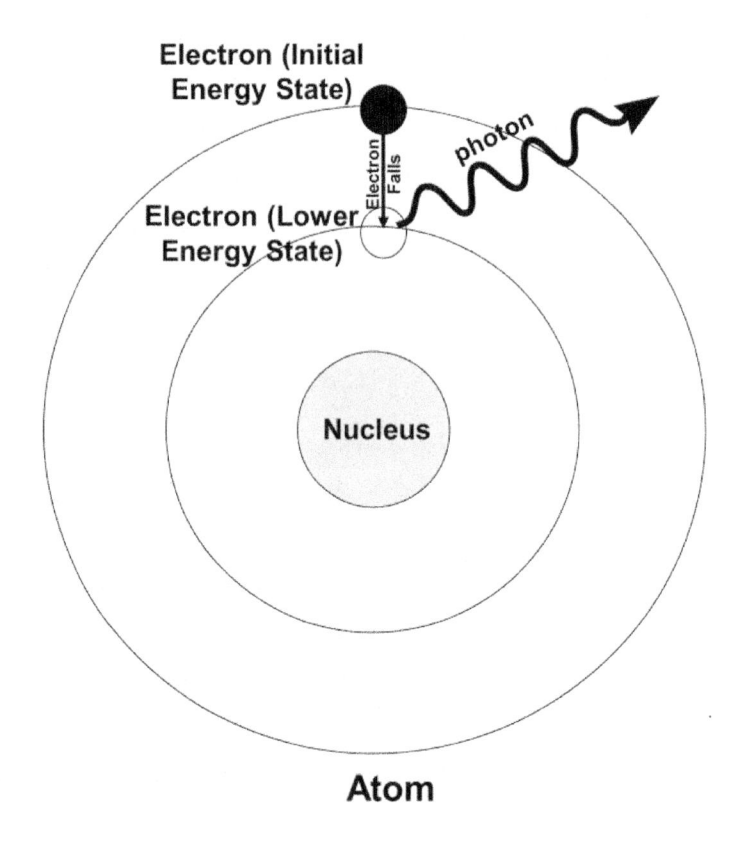

Electron (Initial Energy State)

photon

Electron (Lower Energy State)

Electron Falls

Nucleus

Atom

Relating this to Jesus, the parallels become evident. Let's start with His birth. In His birth, the Divine God is interacting with physical humanity, just like a photon interacts with physical matter. Jesus was sent from God; in fact, He is God. He came to earth as an infant boy, with real human flesh. He took on the form of physical matter, the same physical matter that makes up our own flesh. This birth is pictured by the electromagnetic wave traveling from space and striking the electron in an atom. This electron absorbs the photon, and the electron now exists in a higher energy state, but it is still an electron, and it is still residing in the atom. In this metaphor, the atom represents the earth, and the electron represents a human body.

The electron, which now exists in a higher energy state, represents the incarnation of Jesus on the earth. We know that Mary was a virgin, and that Jesus was supernaturally conceived in Mary's womb without the help of a man. This is perfectly illustrated by the electromagnetic wave interacting with the electron in the atom. When this interaction takes place, it is the photon or particle like nature of light that mediates this process. Notice that the photon has been absorbed by the electron, and the electromagnetic wave is now absent. Also notice that the electron exists in a higher energy state than the other electrons in the atom. This is a picture of the fact that Jesus is fully man, and fully God. The Divine has been wrapped in flesh, and thus, the electron now exists in an elevated energy state. This is illustrated in (1) in the graphic below:

Birth, Death, and Resurrection of Jesus

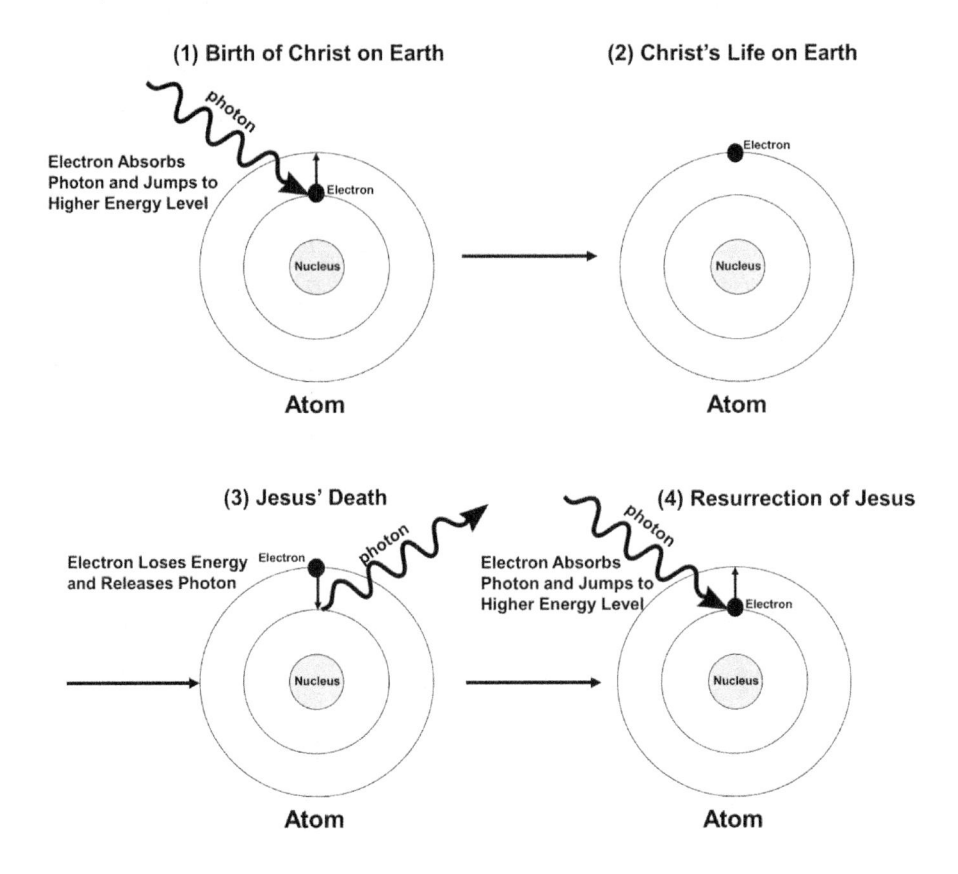

(1) Birth of Christ on Earth

Electron Absorbs Photon and Jumps to Higher Energy Level

photon / Electron / Nucleus

Atom

(2) Christ's Life on Earth

Electron / Nucleus

Atom

(3) Jesus' Death

Electron Loses Energy and Releases Photon

Electron / photon / Nucleus

Atom

(4) Resurrection of Jesus

Electron Absorbs Photon and Jumps to Higher Energy Level

photon / Electron / Nucleus

Atom

Christ lived His life on this earth as fully God, and fully man. Jesus' life on earth is represented as (2) in the figure above, where the electron exists in a higher energy state.

When Christ was put to death on a cross by crucifixion, He physically died, the same way that any human would die. It was a real, physical death in every sense of the word. On the cross, Jesus had taken on the sin of the entire world. In our metaphor, this is represented by the electron falling back down to its original energy level. The electron in this "death" position is related to the physical body of Jesus which is now located in a tomb. Also notice that when the electron "dies", it releases an electromagnetic wave ("light beam") possessing the same exact frequency as the original electromagnetic wave. This light beam represents the Spirit of Christ. Jesus' death is represented by (3) in the figure above. When Christ breathed His last breath on the cross, His Spirit (i.e., the electromagnetic wave) was committed back to the Father:

> Luke 23:46 - "And when Jesus had cried with a loud voice, he said, Father, into thy hands I commend my spirit: and having said thus, he gave up the ghost."

Three days after His death, the scriptures proclaim that Jesus was physically resurrected from the dead. Witnesses, which included the Roman guards guarding His tomb, saw the stone rolled away and the empty tomb. Hundreds of witnesses had physical encounters with Jesus after His resurrection.

> Acts 2:24 - "But God raised him from the dead, freeing him from the agony of death, because it was impossible for death to keep its hold on him."

The resurrection of Jesus from the dead is represented by (4) in the figure above. An electromagnetic wave (photon) interacts with an electron, causing the electron to "resurrect" from its position of death, into a new energy state representing life. Notice that the electron is no longer found in its old energy state. This corresponds to the empty tomb. After the electron is raised from the dead as depicted in (4), the following picture is observed:

The Empty Tomb

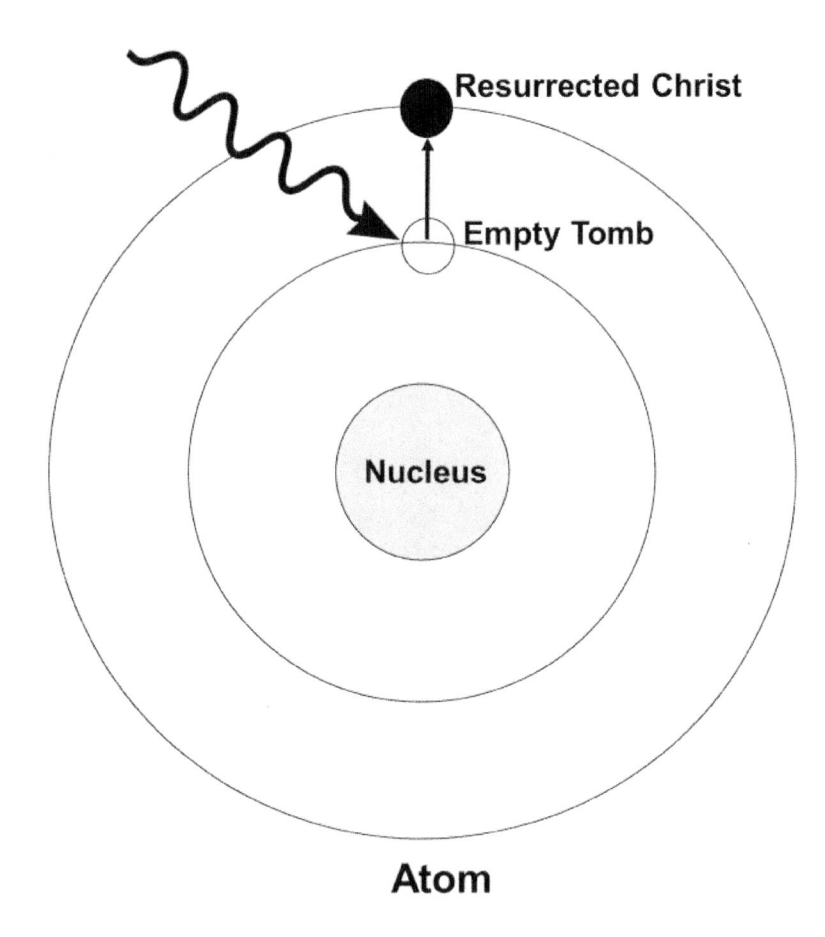

Atom

So, we see a beautiful picture of how the interaction of light with physical matter corresponds with the interaction of God with man; namely in His birth (incarnation), death, and resurrection. The birth, death, and resurrection of Jesus is the central element, the defining feature, and the foundation from which the entire structure of Christianity is built. This truth is what separates the Christian God from all other gods, and we see how the physics of light and subatomic matter interact to tell this story.

Jesus Experienced Birth, Death, and Resurrection ←→ Light Illustrates Jesus' Birth, Death, and Resurrection

> *Matthew 2:1 – "Now after Jesus was born in Bethlehem of Judea in the days of Herod the king, behold, wise men from the east came to Jerusalem, ..."*
>
> *Galatians 4:4 – "But when the fullness of time had come, God sent forth his Son, born of woman, born under the law, ..."*
>
> *John 1:14 – "And the Word became flesh and dwelt among us, and we have seen his glory, glory as of the only Son from the Father, full of grace and truth."*
>
> *Mark 15: 24-26; 37-38 – "And they crucified him. Dividing up his clothes, they cast lots to see what each would get. It was nine in the morning when they crucified him. The written notice of the charge against him read:* THE KING OF THE JEWS. *With a loud cry, Jesus breathed his last. The curtain of the temple was torn in two from top to bottom. And when the centurion, who stood there in front of Jesus, saw how he died, he said, "Surely this man was the Son of God!"*
>
> *Mark 16:6 – "'Don't be alarmed,' he said. 'You are looking for Jesus the Nazarene, who was crucified. He has risen! He is not here. See the place where they laid him.'"*
>
> *I Corinthians 15:3-4 – "For what I received I passed on to you as of first importance: that Christ died for our sins according to the Scriptures, that he was buried, that he was raised on the third day according to the Scriptures."*

XIV. PUTTING IT ALL TOGETHER

Pausing for a moment

Wow, what a journey! Is your mind blown yet? Did you learn something remarkable? The story I just outlined has been a lifelong journey for me, starting with my first introduction to Einstein's Special Theory of Relativity and God's Word as a teenager, progressing through my college and early career years, and finally culminating with the writing of this book in my forty-eighth year of life. The process of writing this book has helped me organize all of these thoughts and to present them in a logical sequence, and I hope it was helpful to structure the material in this way to help you grasp and make sense of the presented concepts.

The thesis presented in this book is the following: God, the Creator, created the laws of physics in such a way as to authenticate His identity and communicate truths about His character and nature. God imprinted His character, nature, and identify onto the physical laws that govern light, space, time, matter, and energy. It is much easier to discount a written text produced by man than it is to discount a message that is embedded in the laws of the universe.

There is only one Person who can tinker around with the laws of physics, and that person is the One who created those laws in the first place: God. It would be quite unusual if God simply created the laws of physics with a sort of meaningless arbitrariness which communicated nothing of significance. In this book, I describe how we can know the identity and nature of the true Creator by studying the hidden messages that He placed into the laws of physics that govern light, space, time, matter, and energy.

By studying science, we learn more about the nature of God. Although the academic establishment professes otherwise, it is my belief that science and faith should not be separated. In fact, they are designed to be intimately connected. The science underlying the natural world is designed to communicate an important message to mankind. By studying science, we learn more about the One who created the laws that science aims to unravel. The laws that govern creation convey a message that is being broadcast throughout the universe twenty-four hours per day, seven days per week. If we separate faith and science, we miss the message and we throw away the real value of science - which is to reveal to us the identity and character of our Creator.

Whether we are studying astrophysics, quantum physics, chemistry, biology, or mathematics, we are learning something new and exciting about God that is not spoken or revealed to us by any other language.

> *Psalm 19:1-4*
> *"The heavens declare the glory of God;*
> *the skies proclaim the work of his hands.*
> *Day after day they pour forth speech;*
> *night after night they reveal knowledge.*
> *They have no speech, they use no words;*

> *no sound is heard from them.*
> *Yet their voice goes out into all the earth,*
> *their words to the ends of the world."*

> *Romans 1:19-20*
> *"....what may be known about God is plain to them,*
> *because God has made it plain to them.*
> *For since the creation of the world*
> *God's invisible qualities—his eternal power and divine nature—*
> *have been clearly seen,*
> *being understood from what has been made,*
> *so that people are without excuse."*

Specifically, we examined the physics of light to determine if God incorporated a message meant for us to understand. This supposition is supported by the fact that God makes special mention of the creation of light in the opening chapter of the Bible:

> *Genesis 1:1-4*
> *"In the beginning God created the heavens and the earth. Now the earth was formless and empty, darkness was over the surface of the deep, and the Spirit of God was hovering over the waters. And God said, "Let there be light," and there was light. God saw that the light was good, and he separated the light from the darkness."*

Throughout the scripture, God relates Himself to light:

> *Daniel 2:22 – "He reveals deep and hidden things; he knows what lies in darkness, and light dwells with him."*
>
> *Isaiah 9:2 - "The people walking in darkness have seen a great light; on those living in the land of deep darkness a light has dawned."*
>
> *I John 1:5 – "This is the message we have heard from him and declare to you: God is light; in him there is no darkness at all."*
>
> *John 8:12 – "When Jesus spoke again to the people, he said, "I am the light of the world. Whoever follows me will never walk in darkness, but will have the light of life."*
>
> *John 1:1-9 – "In the beginning was the Word, and the Word was with God, and the Word was God. He was with God in the beginning. Through him all things were made; without him nothing was made that has been made. In him was life, and that life was the light of all mankind. The light shines in the darkness, and the darkness has not overcome it. There was a man sent from God whose name was John. He came as a witness to testify concerning that light, so that through him all might believe. He himself was not the light; he came only as a witness to the light. The true light that gives light to everyone was coming into the world."*

As we studied the physics of light, a very interesting picture began to emerge. First, we saw that the behavior of light is exceptionally out of the ordinary (to say the least). Light unequivocally does not behave like ordinary objects, and its properties are unique. The most unique of these aspects is light's absolute speed, regardless of the reference frame of the observer. From this distinctive property not shared by physical objects, Einstein derived his Theory of Special

Relativity, which stipulated that space, time, and matter are relative to light's absolute standard. This theory, which obliterated the dogmas of physics, was originally rejected by most mainstream scientists. However, as experiment after experiment was conducted, they each confirmed the predictions made by this theory to many decimal places. Today, Einstein's Theory of Special Relativity is regarded as a scientific fact. The conclusions drawn from this theory mapped perfectly to the characteristics attributed to the Christian God.

Then, we looked closer at the general physics of electromagnetic waves and the wave-particle duality of light. From these principles, a host of other characteristics and unique identifying features of the Christian God were revealed.

It became apparent that light is a physical metaphor for the spiritual nature of God. All of God's qualities are encoded in the physics of light, including both His infinity and humanness in the person of Jesus Christ.

Let's step back and review all of the direct parallels between God and light:

God is Absolute ←→ Light is Absolute

God is Timeless ←→ Light is Timeless

God is Omnipotent ←→ Infinite Power is Required to Reach the Speed of Light

God is Omnipresent ←→ Light is Omnipresent

God Is Omniscient ←→ Light Can Transmit an Infinite Amount of Information

God's Mind is Concealed ←→ Light's Communication is Secret and Unbreakable

God is Self-Sustaining ←→ Light is Self-Sustaining

God is Unified ←→ Light is Unified

God is a Trinity ←→ Light is a Trinity (In Composition)

Father (Invisible, Intangible) ←→ Electric Field (Invisible, Intangible)
Holy Spirit (Invisible, Intangible) ←→ Magnetic Field (Invisible, Intangible)
Jesus (Visible, Tangible) ←→ Photon (Visible, Tangible)

God is a Trinity ←→ Light is a Trinity (In Color)

God is a Trinity ←→ Light is a Trinity (In Spectrum)

Father (Invisible) ←→ Frequencies Below Visible (Invisible)
Holy Spirit (Invisible) ←→ Frequencies Above Visible (Invisible)
Jesus (Visible) ←→ Frequencies Within the Visible (Visible)

Mapping the Trinity onto $E=mc^2$

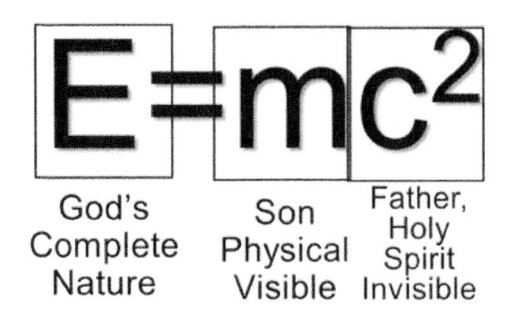

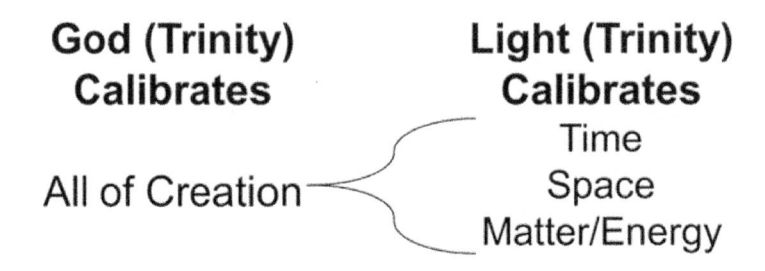

$$E = mc^2$$

E — God's Complete Nature

m — Son, Physical Visible

c² — Father, Holy Spirit, Invisible

God (Trinity) Calibrates — All of Creation

Light (Trinity) Calibrates — Time, Space, Matter/Energy

God Calibrates All Creation ←→ Light Calibrates Time, Space, and Matter/Energy

God Illuminates the Spiritual World ←→ Light Illuminates the Physical World

God is Infinite ←→ Light Has Infinite Frequencies

God Reveals a Portion of Himself to Us ←→ Light Reveals a Portion of Itself to Us

God Holds All Things Together ←→ Light Holds All Things Together

Jesus is Fully God and Fully Man ←→ Light is Fully Wave and Fully Particle

Jesus Experienced Birth, Death, and Resurrection ←→ Light Illustrates Jesus' Birth, Death, and Resurrection

Viewing this entire list of parallels all at one time is quite remarkable. How can the physics of one thing map so perfectly to the divine qualities of the exact God described in the Bible? The same impossibilities we see residing in God are the same impossibilities we witness in light. God's Absoluteness, Timelessness, Omnipotence, Omnipresence, and Omniscience are clearly seen. The same paradoxes we see residing in the God of the Bible are the same paradoxes we see in light. Each member of the Trinity is perfectly described by the physics of light, not only in number but also in attribute. The infinity and humanity of God are all there, in wonderful detail!

XV. WHAT DOES THIS MEAN FOR ME?

There is good news

At the beginning of the book, I told the story about our friends introducing our family to Christianity when I was a teenager. It wasn't too long after that I gave my life to Jesus while I was sitting at a table in a pancake restaurant. At the table, the same friend who had introduced us to the Bible asked me if I wanted to become a believer. I said yes. I prayed with him, and that started my journey with Christ.

I didn't feel fireworks or experience an instantaneous radical transformation, but I did feel like my life had changed at that moment. This marked the beginning of a lifelong transformation that slowly, but surely, changed everything about me. There has been many ups and downs, but even after more than 30 years, I am still being transformed for the better, and I imagine that I will be for a very long time.

So, I ask you the same question: Do you want to give your life to Jesus? I realize that everything I presented in this short book is not 100% proof, but it does make you stop and think, doesn't it? If you weigh this information with all of the other evidence that points to

Jesus being God in the flesh, it paints a very compelling picture that at least deserves our consideration. Before you ask for more proof, let me introduce you to the idea that God is pleased with us when we exercise a certain degree of faith, and this necessarily means that we need to take a leap across a gap that spans where knowledge ends and belief begins. God intends for faith to take center stage:

> Hebrews 11:1,3,6 - *"Now faith is confidence in what we hope for and assurance about what we do not see..... By faith we understand that the universe was formed at God's command, so that what is seen was not made out of what was visible..... And without faith it is impossible to please God, because anyone who comes to him must believe that he exists and that he rewards those who earnestly seek him."*

Now, why should you give your life to Jesus? To understand this, it's first necessary to turn this question around and ask why Jesus first gave His life for you.

The word "sin" is certainly not in vogue these days. Most people would chalk up sin to an antiquated concept that has seen its day. But before we discount the reality and seriousness of sin, let me ask you a question: has there ever been a time in your life where you felt guilt or remorse for doing something that you knew was wrong? Have you ever hurt someone else? Have you ever been selfish? Have you ever looked out for yourself while someone else suffered? Have you ever lied? Have you ever been puffed up with pride and arrogance? Have you ever wasted the time, money, talents, and resources you have been given? Have you ever been unthankful? Have you lived your life without giving God a single thought or any of your time?

I have done ALL of these things, plus many more in good measure. In fact, my list for today isn't looking very good. All of the things I listed above are sin. They take us away from God. And you, too, have sinned. I don't say this to make you feel bad about yourself. We ALL sin. We're all in the same boat, and nobody is better (or worse) than anyone else:

> I John 1:8 - *"If we claim to be without sin, we deceive ourselves and the truth is not in us."*
>
> Romans 3:23 - *"For all have sinned and fall short of the glory of God."*

But why do we sin? Our sin arises from what we are made of. We are made of human flesh, and we are told that our flesh doesn't have the capacity to refrain from sinning. Our flesh does not have the capacity to please God, because it is selfish. We inherited our sinful flesh from the first sinner, Adam:

> Romans 5:12 - *"Therefore, just as sin entered the world through one man, and death through sin, and in this way death came to all people, because all sinned—"*
>
> Galatians 5:17 - *"For the flesh desires what is contrary to the Spirit, and the Spirit what is contrary to the flesh. They are in conflict with each other, so that you are not to do whatever you want."*
>
> Romans 7:18 – *"For I know that good itself does not dwell in me, that is, in my sinful nature (my flesh). For I have the desire to do what is good, but I cannot carry it out."*

So what? Afterall, everyone sins, right? No big deal. Well, there is a problem. There is a consequence to sin. The consequence is death:

> Romans 6:23 – "For the wages of sin is death..."

Unfortunately, the result of sin is death. Eternal death. Our mortality can be directly blamed on our sinful nature. That's pretty bad news. It would seem that our situation is completely hopeless. It is impossible for us to live a life without sin, and as a result we are all destined to surely die. Before we give up hope, though, let's keep reading the rest of that last verse:

> Romans 6:23 – "For the wages of sin is death, but the gift of God is eternal life in Christ Jesus our Lord."

Did you catch that? God provides us with eternal life, and this is a gift that comes directly from Jesus. As we learned in our lesson on light, Jesus is God in the flesh. God left His dwelling to wrap Himself in flesh, and to live on the earth as one of us:

> John 1:14 - "The Word became flesh and made his dwelling among us. We have seen his glory, the glory of the one and only Son, who came from the Father, full of grace and truth."

The fact that the Creator would take on the form of the creation was foretold hundreds of years before it took place:

> *Isaiah 9:6 – "For to us a child is born, to us a son is given, and the government will be on his shoulders. And he will be called Wonderful Counselor, Mighty God, Everlasting Father, Prince of Peace."*

But why did He come? What was His purpose and mission? Was it just to send us greetings and to teach us a few lessons about how to live our life? Well, His purpose in coming was also foretold hundreds of years beforehand:

> *Isaiah 53:4-7 – "Surely he hath borne our griefs, and carried our sorrows: yet we did esteem him stricken, smitten of God, and afflicted. But he was wounded for our transgressions, he was bruised for our iniquities: the chastisement of our peace was upon him; and with his stripes we are healed. All we like sheep have gone astray; we have turned every one to his own way; and the Lord hath laid on him the iniquity of us all. He was oppressed, and he was afflicted, yet he opened not his mouth: he is brought as a lamb to the slaughter, and as a sheep before her shearers is dumb, so he openeth not his mouth."*

These verses record a very sobering message: God took on the form of a human so that He could come to this earth, be rejected by His creation, and suffer and die to pay the penalty for our sin. This was all planned before God even created the universe. Not only did God create us, He also rescues us. Jesus had one mission from the Father, and that mission was to die to pay the price for the sins of all mankind:

> *Romans 5:8 – "But God demonstrates his own love for us in this: While we were still sinners, Christ died for us."*

> *I Peter 2:24 – "He himself bore our sins" in his body on the cross, so that we might die to sins and live for righteousness; "by his wounds you have been healed."*
>
> *I John 2:2 – "He is the atoning sacrifice for our sins, and not only for ours but also for the sins of the whole world."*

After Jesus died on the cross, he was placed in a tomb and a large stone was rolled in place to seal the entrance. Roman guards were positioned to at the entrance to protect His body from being stolen. Even with these safeguards in place, Christ arose from the dead and escaped the tomb.

> *Matthew 28:5-6 – "The angel said to the women, "Do not be afraid, for I know that you are looking for Jesus, who was crucified. He is not here; he has risen, just as he said. Come and see the place where he lay."*
>
> *I Corinthians 15:3-4 – "For what I received I passed on to you as of first importance: that Christ died for our sins according to the Scriptures, that he was buried, that he was raised on the third day according to the Scriptures."*

It's quite incredible to think that the Creator of humanity would choose to become one of His creation and to die in their place, even while His creation rejected Him. Instead of us experiencing death, Christ experienced death. Remember the truth that the consequence of sin is death? Instead of that death being passed on to us, it was passed on to Christ.

What do we need to do in order to receive this wonderful gift?

> John 3:16 – "For God so loved the world that he gave his one and only Son, that whoever believes in him shall not perish but have eternal life."
>
> John 11:25-26 – "Jesus said to her, 'I am the resurrection and the life. The one who believes in me will live, even though they die; and whoever lives by believing in me will never die. Do you believe this?'"
>
> Mark 1:14-15 – "….Jesus went into Galilee, proclaiming the good news of God. 'The time has come,' he said. 'The kingdom of God has come near. Repent and believe the good news!'"

The only thing you need to do is repent and believe the good news!

> Romans 10:9 – "If you declare with your mouth, "Jesus is Lord," and believe in your heart that God raised him from the dead, you will be saved."

Would you like to give your life to Jesus like I did that day while I was sitting in a pancake restaurant? I can personally attest from my experience over the last 30+ years that you will not regret making this decision. There are many decisions you might regret in your life, but this is not one of them. Knowing that your sins are forgiven sets you free to live the life God designed for you and to freely serve the Creator of the universe.

If you are inclined, this is a prayer that you can offer to God:

"God, thank you for creating me! I know that I sin, and I want to turn away from those sins and turn to you as my Lord and Savior. Thank you for

your forgiveness. I believe that you sent Your one and only Son, Jesus, to come to this earth and die to pay the penalty for my sins. I also believe that He was risen from the dead, and that He offers me eternal life. I want to give my life to You, and I desire for You to take control of my life. I am Yours!"

If you said those words and meant them with your heart, or if you offered your own prayer to God, I would love to hear from you. Please let me know that you are a new believer, and I would love to encourage you on your new journey.

Please send me an email to: **DrDougCorrigan@gmail.com**

If you were already a believer before you started reading this book, I would encourage you to ***spread the good news*** by sharing or giving books to others who need to hear the message. This book would also serve as an excellent tool for a small group bible study or devotion. It is also a great educational tool for homeschool families and groups.

I believe that the information in this book can be used by God to draw people to Him. As a young boy, He captured my imagination with this information, and I am sure it will spark the imagination of many others. Afterall, God must have had a very good reason for going through the trouble of engineering the physics of light in such an intricate way that it authenticates His identity and proclaims the gospel. Ask God who He would have you share this message with.

You can get this book at Amazon, or purchase it in bulk at:
www.ScienceWithDrDoug.com

I am also available for speaking engagements (face-to-face, or virtually online). If you would like me to speak to your group about this topic, please email me at:
DrDougCorrigan@gmail.com

The Author of Light

Printed in Great Britain
by Amazon